# Advances in Global Change Research

## Volume 72

More information about this series at https://link.springer.com/bookseries/5588

Lia Marchi

# Designing Sustainable Factories

## A Toolkit for the Assessment and Mitigation of Impact on the Landscape

Springer

Lia Marchi
Department of Architecture
University of Bologna
Bologna, Italy

ISSN 1574-0919 ISSN 2215-1621 (electronic)
Advances in Global Change Research
ISBN 978-3-030-92226-9 ISBN 978-3-030-92227-6 (eBook)
https://doi.org/10.1007/978-3-030-92227-6



This Springer imprint is published by the registered company Springer Nature Switzerland AG
The registered company address is: Gewerbestrasse 11, 6330 Cham, Switzerland

# Foreword

About six years ago, Orogel, an Italian leading agri-food company, engaged our Department of Architecture for support in mitigating the impact into the landscape of their new automatic refrigerating warehouse: a huge parallelepiped of 60 per 120m, 30m high, whose construction was beginning at that time in Cesena (Italy).

Orogel's major concern was to reduce the visual impact of the new facility, as part of the company's strategic social engagement program to promote the culture of healthy food and its sustainable production, which had already led to adopt innovative technologies both in building the facilities and in working processes with the aim of reducing the environmental impact of production.

Together with my colleague Valentina Orioli, Professor of Urban Planning, I decided to accept this multidisciplinary challenge. Thus we were appointed supervisors of a 3-year Ph.D. research on this topic, which Orogel supported with a generous scholarship. This is how Lia Marchi's Ph.D. project began, whose results have made it possible to write this book.

This gave us the opportunity to address the issue of assessing the impact of industrial plants on the environment, which, despite being an emerging topic in both theory and practice, still lacks a structured body of knowledge and systematic understanding. Hence the research has tackled the issue by adopting a holistic approach, which includes within the surveyed impacts also the perceptual and visual disturbances of factories on the landscape. So, the assessment was extended beyond the environmental physical matrices that are usually the only considered by almost all current protocols.

Since the early stages, Lia Marchi has driven her research looking at the topic by a wider scope and vision than the aim of mitigating only the impacts of a single company's facility. Therefore, rather than starting from a specific case, the study has seeked to produce a generally applicable outcome. This is, namely, a toolkit that can be adopted to assess and mitigating the impact of a large variety of industrial buildings, which can differ by location, function, size, and shape. This pushed the adoption of a broader and holistic perspective regarding the notion of landscape compatibility too: instead of focusing on the sole aesthetic aspect, the environmental, social, and perceptual-visual interferences of manufacturing sites with the landscape have been integrated into a coherent framework.

The finding that there are powerful synergies between the different dimensions of the landscape, as shared by many authors, justifies this choice, which also appears as appropriate when considering the effect that can be produced by integrating mitigation measures belonging to the aesthetic and environmental spheres, instead of simply combining single sectorial actions. Especially small- and medium-size businesses could benefit from this enlarged perspective, as they need to go beyond aesthetic enhancements of the corporate image to incorporate real sustainable practices into their factories and processes. While green washing often represents an additional cost to the business, as this book well illustrates, a holistic approach to sustainable buildings can generate great benefits for both the construction and operation of the factory if it is adopted since the beginning of the design process.

In order to move along this pathway, entrepreneurs must have a clear understanding of the direct and indirect, tangible and intangible benefits that can be obtained from a low-impact factory, in terms of both costs and emissions savings thanks to efficient systems, and increased competitiveness through the improvement of the corporate image and customer appreciation. This book gives all of them good arguments and suitable tools to do that, such as the large collection of carefully documented examples of low-impact factories and the precious repertoire of mitigation tactics extracted from the best practices found in the former. The diverse scales and facets of the landscape are coherently brought within a comprehensive but flexible assessment structure, grounded in the established procedures and methodologies of the U.S. GBC LEED Green Building Rating System, which allows a proper multi-criteria approach. As a result, a toolkit to assess and mitigate the impact of factories on the landscape in a comprehensive manner is made available.

The outcome is a multidisciplinary study that spans from considering environmental performance of factories at the building level to the landscape quality of the site, and even beyond by addressing territorial promotion.

As much as experimental, it comes of a handily and inspiring tool for companies and designers who want to embark on this trip, which appears as much challenging as necessary.

With her excellent "Designing sustainable factories", Lia Marchi demonstrates that it is time to provide more sustainable sites and more sustainable industrial processes, and that not only is this necessary but also possible and even convenient.

Ernesto Antonini
Professor of Technology for Architecture
University of Bologna
Bologna, Italy

# A Green Factory for a Sustainable Business

For over 50 years, we have been cultivating the land, working the soil, seeding, and harvesting the best vegetables to freeze: we take the best from each season. We cultivate in a sustainable way adopting many good practices. We freeze using the most advanced technology, never forgetting that our mission is to provide healthy and high-quality products that are also sustainable.

We have improved our processes so that technology helped us be more environmentally friendly. We have reduced the consumption and impact of plastic, reused energy and water, reduced emissions into the atmosphere, and built biogas plants powered by production waste so that nothing was wasted.

We have adopted a sustainable approach also addressing the places where we work and their people, from our farmers to the factory workers, because people too are protagonists of the ecosystem to which we belong. We are always connected with what surrounds us.

We want to be "Sustainable for Nature".

A great vision that needed to be completed. We realized that our commitment to sustainability had to extend to the design and construction of our manufacturing facilities, minimizing their impact, and, most importantly, harmonizing their massive volumes in the landscape so that they would not appear decontextualized or conflicting. Indeed, the conviction arose that the factory should not only be technologically advanced, efficient, low-energy, safe, and profitable for our business, but also have a pleasant, attractive image, and be "green" in the eyes of the local community too, as an act of social responsibility. This was the starting point for the new Orogel 3 site, which was built in addition to our main facility in Cesena (Italy).

Because this was a new thought for us, we needed new ideas. We had to find them in someone with a fresher, more contemporary perspective. Thus, we decided to fund a doctoral research project to go deeper into the subject and suggest a novel design approach to our industrial architecture.

The study validated our assumptions by strengthening our belief that a green and attractive factory is not only a value for all, but also a profit factor for our competitiveness. Now it is up to us to put the results of this research into practice. The outcomes illustrated in this book will serve as our guideline, but we hope it will also serve all firms aspiring to be 360° sustainable. Another step further in our social responsibility: starting from our needs, we hope that this toolkit can be useful to other companies all around the world to become green.

Valter Zino
Technical Director at Orogel
Italy

# The Landscape Challenge, Balancing Protection with Transformation

The book examines the impact of industrial settlements on the landscape, which is a highly relevant topic nowadays. This work is the outcome of a Ph.D. research project aimed at investigating the Emilia-Romagna context (Italy) and proposing strategies for the landscape mitigation of food and beverage production facilities, which are often near rural areas, due to their function. Because of the significance of the topics addressed, the paucity of literature and case studies, as well as the experimental nature of the assessment tools developed, the study made use of a wide range of references from the international scene. As a result, the research assumed the breadth and the character of "universality" that can be appreciated in this book.

However, in this shift from a study of the local situation to an approach that can be implemented on a global scale, some aspects remain very contextual because they are dependent on the conception of territorial governance system and landscape protection tools. Even if the concept of landscape is now widely shared at international level, the way in which actions on protection are taken is unique to each country—whether through specific procedures or urban planning tools. Thus, it represents an element with different technical and cultural contents, with which the proposed toolkit must necessarily deal. The Italian situation is indeed quite peculiar.

"Paesaggio" (landscape) is a polysemic term that refers to a variety of meanings. Without going over its various definitions—which is a fairly common exercise in the scientific literature as proven by the wide range of references that can be found in—we will limit ourselves to observe that the word "paysage" originally referred to a pictorial genre, and thus to natural scenarios experienced in aesthetical terms. Later, the survey of landscape was enriched with multiple references to the objective reality, up to the point of connoting it as an ensemble of natural spaces and living communities, but also as a product and heritage of the work of man in the territory. These two latest connotations are certainly more prevalent in the Anglo-Saxon word for landscape, which "contains the ground" (i.e., land).

The geographer Lucio Gambi's entry "Paesaggio" for the *Enciclopedia Italiana* (Italian encyclopaedia) provides an effective synthesis of the various ideas of landscape and the meanings it has assumed over time in relation to different disciplinary approaches, as well as a definition of landscape as a product of a cultural process rather than a specific natural situation, which perfectly fits with the Italian case:

> The canvas, sometimes very thick, sometimes thinner, laying like a mantle on the planet; which is the result of the work of hundreds of centuries [...] and which has to be judged for the actual benefits or losses that it brings to the society, and for the nature and the extent of its relations with the elements of the environment (Gambi, 2000).

In fact, this view of the landscape as a "palimpsest" (Corboz, 1985), with all the implications for man's responsibility as a "landscape maker", is very different from the approach that has been consolidated in the practice of the landscape protection since the first national laws at the beginning of the twentieth century.

In Italy, the first measure in terms of landscape protection is the Law no. 441/1905, commissioned by the Minister of Agriculture Luigi Rava, together with the Florentine deputy Giovanni Rosadi: the subject was the inalienability of the coastal wrecks of Ravenna's pine forest. In the following years, Rosadi, who was very active in this field, promoted other measures to protect specific landscape assets, until the first systematic measure, which was pursued by Benedetto Croce (Minister of Public Education), was approved: Law no. 778, 11 June 1922, "For the protection of natural beauties and properties of particular historical interest". This regulation defines the landscape to be protected as "permanent asset whose conservation presents a significant public interest due to their natural beauty". While the tool proposed for their conservation is essentially the isolation from the context, both physical and normative; the perimeter that imposes the constraint or the *ad hoc* rule; or the creation of a special area of protection, i.e., the Park.[1]

The same aesthetic perspective is recognizable in the following Law No. 1497, 29 June 1938 "Protection of natural beauties" that institutionalized a "passive" approach to protection for a long time, since it was fundamentally based on environmental restrictions, selected on the basis of aesthetic qualities, partial in the extent of application, and decoupled from urban planning regulations.

Unfortunately, the separation of urban planning regulations and landscape protection law is a permanent feature of the Italian context, where the two subjects are still managed by different authorities: according to the Italian Constitution (Article 117), territorial government is a matter of concurrent legislation, and, despite the Law no. 1150, 17 August 1942 "Urban planning Law" remaining in force, there is now a wide regional autonomy. Instead, the protection of the environment, of the ecosystem and cultural heritage is the exclusive power of the State. Although the Article 9 of the Constitution mentions the relevance of the landscape to be safeguarded as a national heritage, there is no reference of this word in the discipline of exclusive and concurrent legislation.

---

[1] The first national parks in Italy are Gran Paradiso (1922) and Parco Nazionale d'Abruzzo (1923). The national parks of Circeo and of Stelvio date back to the following decade.

Despite this separation, which could be partially overcome in the years to come with the effective development of landscape plans on regional scale with the cooperation of the Ministries of Culture and of Ecological Transition,[2] the landscape protection legislation has evolved over time, partially recovering the original flaws.

The concept of protection has definitively surpassed the aesthetic perspective with the so-called "Decreto Galasso", converted to the Law no. 431 on the 8 August 1985, "Protection of areas of particular environmental interest", which applies to specific portions of the territory identified for their geographical and environmental peculiarities. Through this law, an interpretation of the landscape as a "structural" element of the territory to be safeguarded by the imposition of environmental restrictions was introduced into our legal system. The Galasso decree also included a mandate to the Regions to integrate landscape protection measures into territorial planning, either by including them in the Regional Territorial Plan or by developing *ad hoc* instruments (Ciccone, 1993).

The Regional Territorial Plan of Emilia-Romagna, adopted in 1989, belongs to this "first generation" of tools.

This plan represented an attempt to adapt the law's approach by incorporating protected areas into a comprehensive interpretation: 23 "landscape units" were distinguished, and within these "systems", "zones", and "elements" that features the various regional landscapes. However, the prevailing approach in the Plan implementation was still restrictions-based. So, even though protection has now been extended to the most vulnerable parts of the regional territory, the selectivity and partiality as critical element of the approach have not been entirely overcome. The "constraint" assignment has been confirmed as the basic means to control protected perimeters (Bottino, 1993).

In this framework, the European Landscape Convention—adopted by the Committee of Ministers of the Council of Europe in Strasbourg on 19 July 2000 and open for signature by Member States in Florence on 20 October of the same year—marks a cultural turning point in the approach to landscape protection. In fact, the Convention affirms the need for a new perspective on the landscape, allowing an effectively "integrated" interpretation that considers not only the physical data and how it has been defined over time, but also the fundamental perceptual and relational dimension that characterize all landscapes. According to the Convention:

> "Landscape" means an area, as perceived by people, whose character is the result of the action and interaction of natural and/or human factors (Council of Europe, 2000, Art. 1).

It also legally recognizes the landscape as an asset, regardless of the value that is actually attributed to it,

> as an essential component of people's surroundings, an expression of the diversity of their shared cultural and natural heritage, and a foundation of their identity (Council of Europe, 2000, Art. 5).

---

[2] According to what established in the Article 143 of the Code of Cultural Heritage and Landscape (D.L. 22 January 2004, no. 42).

On the practical level, this conceptual shift has had a significant impact. The Convention established that, because the landscape must be treated as an asset regardless of its value, the *entire territory is landscape*. In other words, the entire European territory *has* or *could have* landscape relevance. This value is based on the perceptual relation that populations establish with their living environment: that is, based on a subjective and relational component as crucial part of individual well-being and local identity, rather than on a value of the landscape itself. According to the Convention,

> the landscape contributes to the formation of local cultures and that it is a basic component of the European natural and cultural heritage, contributing to human well-being and consolidation of the European identity; [...] the landscape is an important part of the quality of life for people everywhere: in urban areas and in the countryside, in degraded areas as well as in areas of high quality, in areas recognised as being of outstanding beauty as well as everyday areas (Council of Europe, 2000, Preamble)

In Italy, the Convention has represented a key reference in both the theoretical debate and the planning and design practices related with the landscape. Thanks to it, in fact, it has been highlighted

> the necessity of moving from a sectoral and passive policy to one that is integrated, intersectoral, active, and participatory. The European Landscape Convention expresses – even if in a open and challenging way – this new patrimonial, community, and socially regulated sense of the landscape value[3] (Poli 2012, XXXI).

On this basis, the Code of Cultural Heritage and Landscape has introduced a new definition of landscape, such as

> territory expressive of identity, whose character originates from the action of natural and human forces, and their interactions, advocating their conservation in connection to those aspects and characters that comprise a material and visible manifestation of national identity, as an expression of cultural values[4] [Art. 131]

On the one hand, this approach is reflected in the development of new Landscape Plans, which is proposed as a cooperative activity between the State and the Regions in order to integrate the landscape theme into territorial planning.[5] On the other hand, it refers to a broader cultural dimension, which encourages the adoption of active and participatory policies by citizens.

The aspiration for an integrated and participatory approach to landscape protection and enhancement is evident in the "second generation" of Landscape Plans, such

---

[3] The original text in Italian is: "la necessità di passare da una politica settoriale e passiva a una politica integrata, intersettoriale, attiva e partecipata. La Convenzione Europea del Paesaggio incarna—pur in modo aperto e problematico—questo nuovo sentimento patrimoniale, comunitario e socialmente controllato dei valori paesaggistici".

[4] The original text in Italian is: "territorio espressivo di identità, il cui carattere deriva dall'azione di fattori naturali, umani e dalle loro interrelazioni, proponendone la tutela relativamente a quegli aspetti e caratteri che costituiscono rappresentazione materiale e visibile dell'identità nazionale, in quanto espressione di valori culturali".

[5] Since the adoption of the Code in 2004 only 4 Italian Regions have approved their "Piano Paesaggstico Regionale": Tuscany (2015), Puglia (2015), Piedmont (2017), and Friuli Venezia Giulia (2018).

as the PPTR adopted in 2013 by Puglia.[6] This represented an attempt of "operative translation" of the Convention's principles. In this plan, the perceptual dimension and design approach are put in synergy aiming to underline the identity and connotative component of the landscape.[7] On the one hand, a plan structured and conceived in this manner reminds us of the agricultural landscape's centrality in terms of the overall image and identity of the territory; on the other hand, it emphasizes the importance of applying a careful and sensitive approach to every scale of the project, paying greater attention to the way the landscape is actually treated in the local dimension and through specific transformations.

These considerations must be assumed not only by the people who "live" (in) the landscape each day, but also by the designers who work "in" the landscape, so that their projects are increasingly sensitive to landscape values. This not only from an environmental point of view, but also from that of identity—thus involving the perceptive and aesthetic dimensions—which by now also the territorial planning recognizes as crucial in *each* landscape.

The research that is presented in the book tries to embed this vision, and to adapt it in an operative manner to a particularly complex topic in the Italian context, which fits perfectly into the challenge of conceiving the transition not only as an environmental and economic effort, but also as a cultural project.[8]

Valentina Orioli
Professor of Urban Planning
University of Bologna, Bologna, Italy

## References

Bottino, F. (1993), Il piano paesistico dell'Emilia-Romagna, in Campos Venuti, G., Oliva, F. (eds), *Cinquant'anni di urbanistica in Italia. 1942–1992*, Laterza, Roma-Bari, pp. 166–178.

Ciccone, F. (1993), La proposta originale dei piani paesistici, in Campos Venuti, G., Oliva, F. (eds), *Cinquant'anni di urbanistica in Italia. 1942–1992*, Laterza, Roma-Bari, pp. 150–165.

Corboz, A. (1985), Il territorio come palinsesto, *Casabella*, n. 516, pp. 22–27.

Council of Europe (2000). *European Landscape Convention*, http://www.convenzioneeuropeapaesaggio.beniculturali.it/uploads/Council%20of%20Europe%20-%20European%20Landscape%20Convention.pdf.

Gambi, L. (2000). *Paesaggio*, in: Istituto della Enciclopedia Italiana, *Enciclopedia italiana di scienze, lettere ed arti*, Tomo II, pp. 395–96.

Poli, D. (2012). *La 'riemersione' del paesaggio nel nuovo Piano di Indirizzo Territoriale della Regione Toscana*, in D. Poli (eds), *Regole e progetti per il paesaggio. Verso il nuovo piano paesaggistico della Toscana*, Firenze University Press, Firenze, pp. XXVII–XL.

---

6 The Landscape Regional Plan of Puglia was adopted in 2013 and finally approved in 2015. http://www.sit.puglia.it/portal/portale_pianificazione_regionale/Piano+Paesaggistico+Territoriale/Documenti/PPTRAdottato/Portlet+Visualizza+PPTRWindow?azionelink=tornaHome&action=2 (accessed on: 10 October 2021).

7 The reference is in particular to *Progetti Integrati di Paesaggio sperimentali* and documents *I Paesaggi della Puglia* and *Laudatio Imaginis Apuliae*.

8 Ursula Von der Leyen about the New European Bauhaus, https://europa.eu/new-european-bauhaus/index_en (accessed on: 10 October 2021).

# Contents

# Abbreviations

| | |
|---|---|
| ELC | European Landscape Convention |
| EU | European Union |
| GBC | Green Building Council |
| GBRSs | Green Building Rating Systems |
| GHG | Greenhouse Gas |
| HVAC | Heating, Ventilation, and Air Conditioning |
| NC | New Construction |
| NEB | New European Bauhaus |
| PA | Perceptual-aesthetic aspects |
| R | Renovation/retrofitting of existing building |
| SDGs | Sustainable Development Goals |
| SMEs | Small and Medium Enterprises |
| VOC | Volatile Organic Compounds |

# List of Figures

# Chapter 1
# The Impact of Industrial Facilities on the Landscape

**Abstract** Whether it comes to a small warehouse or a vast industrial area, the impact of factories on a population's everyday landscape is complex and often adverse. The effects can cover an array of interrelated scales and facets, ranging from environmental concerns to perceptual-aesthetic and social issues, at site, building, and process levels. Location, site quality, technological and visual aspects, function, and resource management are all cross-sectional subjects relevant to the design of a sustainable factory. For instance, colours, materials, and the morphological features of the open space in a manufacturing site can affect a number of aspects, from the microclimate and the attractiveness of the site, to the well-being of its employees. Now global challenges require environmental, social, and "visual" interferences of a factory within a landscape to be eliminated, mitigated, or at least compensated for. Outsourcing the negative effects as in the past is no longer acceptable, nor is the consideration of only environmental impacts. Beyond a hesitant change of mind by entrepreneurs, a robust and expanding policy framework is propelling this transition; the European Green Deal and the New European Bauhaus, in particular, are driving this change in Europe. This chapter concludes that a more holistic approach to sustainability is needed when designing a new factory or retrofitting an existing one.

**Keywords** SME · European landscape convention · Environmental impact · Visual pollution · Worker's well-being

Imagine a world where factories clean the water they use, generate more energy than they require, protect biodiversity, utilise sun and wind to ensure the comfort of their workers, provide lovely green spaces, climbing walls, and skate parks on their roofs. Imagine riding your bike to work through a beautiful landscape of industries, offices, and houses in harmony with the nature around them. This ideal world should not simply be a future to aspire towards, but something that can inspire practices to be adopted in the present.

Although many seeds of change have been planted during the preceding decades, the situation is still far from ideal. On the one hand, industry has improved people's quality of life by providing job opportunities, prosperity, and well-being. On the other hand, it has damaged our living environment, often radically altering the balance

 
L. Marchi, *Designing Sustainable Factories*, Advances in Global Change Research 72,
https://doi.org/10.1007/978-3-030-92227-6_1

of natural and social ecosystems. Plenty of books have been written about how money has established states and destroyed them, opened new horizons and enslaved millions of people, set the wheels of industry in motion and, meanwhile, contributed to the extinction of countless species (Harari, 2011). For many years, profitability and functional requirements have been the only guiding forces behind the design and construction of factories, up to the point that industrial sites have nearly always been seen as detractors of a landscape (Bennet, 1952; Cassatella & Gambino, 2013; Provincia di Milano, 2012). They are places where there is little if any room for sustainability, beauty, and the needs of a community.

As a result, whether it concerns small or large sites, family-run businesses or multinational corporations, urban plans or spontaneous sprawl, the impact of factories on the living environment is always complex, multifaceted, and often conflicting.

First, industrial sites themselves are complex. Spaces of production have diverse forms and dimensions, and their names vary accordingly: factory, plant, facility, complex, production site, industrial area, manufacturing centre. Recently, warehouse, data centre, and logistics platform have also been added. In general terms, this book will address every construction that relates to manufacturing[1] (Eurostat, 2008). This includes, among others: food and beverage, textile, chemical, and metal sectors. Because each category has its own set of needs and requirements, the specialisations of the respective buildings have substantially increased over the centuries. This has strongly influenced the design of production sites, whose form simply follows their function.

Beyond the influence of function, existing industrial complexes can be the result of planned urban zoning, thus being shaped as specialised areas of the city under particular design criteria, or they might originate from autonomous private initiatives. In general terms, planned zoning results in homogeneous and regular constructions, even if they are not necessarily synonymous with high-quality projects, while private initiatives often produce more disparate constructions. The Italian 'Veneto case' is an illustrative example of the latter, where in few decades the landscape had been invaded by factories of every shape and size as a consequence of a chaotic and uncontrolled process of territorial sprawl (Boeri et al., 2013; Cipriani, 2012). This represented an act of "destruction" rather than a "transformation" of the Italian landscape (Turri, 1979).

Complexity is inherent in their different sizes and quantities too. Taking Europe as an example, the industrial sector was made up of over 2 million companies in 2017: 96% of which were small companies, 3.4% medium-sized, and less than 1% large (Eurostat, 2020a). Since most European Enterprises are Small and Medium-sized (SMEs), it follows that the whole continent must be invaded by a multitude of relatively "little" constructions devoted to hosting production activities and functions. Italy is the second country in Europe—after Germany—for the number of buildings classified as 'industrial', where they account for a surface area of over 500 million square meters (Ciaramella & Puglisi, 2016).

[1] Category C—Manufacturing by NACE_R2.

Thus, it is clear that the types of interference with the landscape and the types of intervention that can be implemented on such a diverse and vast scale are difficult to generalise.

Because of the dimension of industrial sites, a second issue to address concerns the many scales at which the interferences occur—a further element of variability that increases the complexity of the topic. At the *site* level, possible detrimental effects deal with the relationships between the factory and its surroundings, the quality of the infrastructure, open space, and natural environment, as well as the visual perception of the complex, and the services it provides both to its employees and the local community. Interferences with the context at the *building* level mainly pertain to the factory architecture, its morphological, technological, and functional features. This includes, for example, the energy performance of the building in relation to its shape, the behavioural patterns of the employees, the operation of its systems, and the properties of the building's envelope. At the *process* level the consequences of the interaction between the building and its systems are typically considered in terms of resource and energy use and efficiency, thus their effects on the local environment are taken into account.[2]

Despite frequently being examined separately, the three levels are inextricably linked, and the same design decision may even simultaneously affect all of them. For example, the energy output from the process (typically as thermal energy) can impact the building's interior comfort through overheating (i.e. an unwanted result) or, on the contrary, it might be recovered to warm the facility and therefore save on the use of additional energy (i.e. a positive effect) (Ball et al., 2013). The colour of the façade provides another interesting example: it pertains directly to the building level as it features its aesthetic and energy performances, but it also influences the perception of the factory in the landscape and the outdoor space's microclimate.

The last example assists in the introduction of the third issue, as it emphasises the complex relationship between factories and their surroundings: that is, the landscape is a multifaceted and broad concept involving several dimensions and interconnected components. In fact, since the European Landscape Convention (ELC) extended its notion to consider natural and human factors as diverse but integrated facets of the territory (European Landscape Convention, 2000), we are expected to regard these dimensions as equally important when defining the quality of a place. More specifically, the Landscape Character Assessment[3] (Tudor, 2014) identifies three complementary dimensions of the landscape that should be considered: (i) environmental, (ii) aesthetic-perceptual, and (iii) social/cultural. Therefore, throughout the following dissertation, the term 'landscape' will refer to this complex entity, and is not to be considered as a synonym of picturesque scenery or a panorama as is commonly the case (Settis, 2010). As a consequence, the assumption that all territory is landscape as people's living context, introduced by the ELC, has prompted

[2] As the book focuses on architecture of factories rather than manufacturing in general, the Author uses the term 'process level' mainly referring to the interactions of the production stage with the building and its technical systems.

[3] LCA is the reference document in Europe for the landscape assessment (Tolli et al., 2016).

the consideration that it is no longer possible to divide between scenery to maintain, conserve, or discard (Allegri, 2015). Ordinary landscapes, industrial ones included, deserve attention because their quality affects the identities of its inhabitants.

The *environmental* impact of factories on the landscape involves a series of effects on the ecosystem's physical matrix. A manufacturing site can for example negatively affect the quality of air, soil, water, material resources, flora and fauna. On the *perceptual-aesthetic* level, very strong contrasts with the scenery are frequently recorded, in terms of lines and forms, colour, texture, sounds, smells, touch/feel. Intangible values can also be affected, such as contrasts with local inhabitants' memories and emotional attachments to the place, and an interruption in their preferences and habits. These in turn might cause *social* disturbances in the building's employees and the local community, affecting their psychophysical health and well-being, and frequently disrupting their sense of place and the territorial value of the area.

As an example, take a small community wedged between the city and the countryside, where residents have chosen to settle for the peaceful view on the hills, the quality of air, the proximity to the forest and its fauna. One day, a company decides to settle its big, grey-smoking, anonymous facility in front of their dwellings, in place of a children's playground. It is quite simple to imagine the kind of emotions that this will trigger in the residents of that community: anger, concern for their health, disappointment, dissatisfaction with the diminished value of their home, and so on. These are only a small number of the effects on the local inhabitants; generally speaking, many other detrimental effects could be added.

On these premises, the first question to answer is this: what kinds of adverse effects do factories currently have on people's everyday landscape?

## Types of Impact

The following pages will outline the most common interferences of factories by adopting a perspective that emphasises links among various scales and facets of the landscape. The impacts are organised by cross-cutting topics, some of which are the typical evaluation categories of green buildings. Table 1 reports the most frequent negative effects of industrial areas as found in both the scientific and technical literature.[4]

**Location**. The location of a building is as important as the way in which it is built since a significant share of emissions and costs are associated with transportation (U.S. GBC, 2014). In fact, according to Reichardt (2015), selecting the right place is the most "far-reaching" decision one could take when building a new factory. Because, typically, the further the area is from a city, the cheaper the lot of land, companies often select sites that are distant from residential settlements

[4] The examined sources include studies about the environmental compatibility of factories, on their visual impact or social disturbances on neighbours; guidelines and regulations at different policy levels; activities, project and sector reports on factories' environmental sustainability and landscape quality.

**Table 1** Recurrent impacts of factories on the landscape and its primary affected dimensions

| Impact | Landscape aspect | | |
|---|---|---|---|
| | ENV | PA | SOC |
| Soil consumption | • | | |
| Increased traffic flow and atmospheric emissions | • | | |
| Spatial disorder | | • | |
| Environmental fragmentation | • | | |
| Reduction of ground permeability | • | | |
| Decontextualization | | | • |
| Fragmentation of the scenery | | • | |
| Visual disorder | | • | |
| Lack of outdoor quality spaces | | | • |
| Noise pollution | | • | |
| Mono-functionality | | | • |
| Energy consumption | • | | |
| Hygro-thermal discomfort | | | • |
| VOC emission | | | • |
| Low design quality and visual detraction | | • | |
| Lack of indoor social spaces | | | • |
| Groundwater pollution | • | | |
| GHG emissions | • | | |

Legend: Environmental (ENV)/Perceptual-aesthetic (PA)/Social (SOC)

and/or from major transportation routes. However, it is this decision-making that causes the number of those who can reach their workplace by bike or public transport to decrease, while causing the transportation costs of both people and goods to increase, as does the use of non-renewable resources (i.e. fossil fuel), related Greenhouse Gas (GHG) emissions, and traffic jams (Cipriani, 2012; Gandolfi, 2011).

The setting of a factory in a certain location can affect its visual and perceptive quality too. In particular, technicians refer to 'urban disorder' when the site is the result of private initiative, autonomous and uncoordinated, and scattered throughout the territory (Busquets i Fabregas, 2007; Cipriani, 2012; O.P.U.Q.P., 2015). For instance, in a rural context, industrial sites frequently create a degree of visual fragmentation in the continuity of the scene: obstruction of visual channels along principal routes, disruption with the physical pattern of the landscape, and unexpected scale change are some of the resulting effects. In general, the more natural the environment, the greater the contrast with the factory. But while it is unacceptable to design a bad factory, an industrial area allows less strict mitigation measures to be considered during the design stage.

The same is true for light and noise pollutions (Busquets i Fabregas, 2007). The first is particularly relevant in rural settings; the second is more concerning in nearby residential neighbourhoods during the daytime due to large traffic flows and intense operations within service areas.

**Site quality and technological features**. The architecture of the building and its relationships with the landscape, as well as the technical choices, affect both the perceptual-aesthetic and environmental quality of the area. Some features are essential to the building type and cannot be changed: industrial sites generally take up large portions of land because manufacturing logistics necessitates horizontal development over height expansion. Thus, facilities rarely exceed two storeys, otherwise the handling of goods becomes problematic. As a result, factories are always very large constructions compared to their surroundings. However, there are some design variables that contribute to making a bigger impact.

For example, the site layout (i.e. the arrangement of buildings and open spaces within the area) and the building shape are commonly out of harmony with the scenic patterns, lines, features, and landforms, causing visual and perceptual disorder (O.P.U.Q.P., 2015). In addition, the outdoor space is frequently characterized by hardscape, such as large impervious ground for parking, loading and unloading areas, and roads, all of which contribute to the worsening of the heat island effect and water runoff, as do the dark bitumen roofs and the highly absorbing materials largely used for the building envelope.

**Spaces and services for people**. In turn, the lack of relation between the site and its surroundings, between the building and its open spaces, the quality of the open spaces themselves, and of interior areas and services, have an impact on the liveability and attractiveness of the site. Productive complexes are frequently devoid of outdoor leisure spaces, green areas, and pedestrian paths. Even when some attention is devoted to these in the design stage, they are usually treated only as residual spaces in the shadow of gigantic buildings (Busquets i Fabregas, 2007; Lombardia, 2011). As these sites are generally fully dedicated to work (mono-functional), somehow companies feel allowed to neglect quality services in the area, and even social areas indoors, such as relaxation rooms and cafés. All of this detracts from the well-being of employees and local residents—who will not easily accept a new factory disturbing their landscape without receiving some positive benefits as compensation (Herrmann et al., 2015).

**Constructive features**. The technological features of the building not only influence its energy demands and resource consumption, but also the comfort, health, and productivity of its employees. The vast majority of existing facilities have poor energy performance due to their design and construction techniques. Usually they are prefabricated buildings, with inadequately thermally insulating envelopes, careless of the local microclimate, and equipped with obsolete and inefficient technical installations (Delpiano & Trombetti, 2011; Veglia & Bono, 2007). As a result, building operations are responsible for a significant share of pollutants and GHG emissions, and the same is true for the energy embedded in components and materials.

This affects on the indoor air quality and thermo-hygrometric comfort of the occupants. In some cases, harmful levels of Volatile Organic Compounds (VOCs) from construction materials and indoor noise pollution (e.g. low-frequency machinery) are registered, which might have a negative influence on the long-term health of the workers (Provincia di Bologna—Servizio Artigianato Commercio e Industria, 2009).

**Aesthetic features**. The aesthetic features of the site and the building have an impact on who works in the building, on the community, as well as on the territorial value and the company's image. Rather than outstanding, memorable, pleasant places of production, factories are frequently perceived as visual detractors due to their massive shape, contrasting colour, and disordered layout. They are characterised by banal, repetitive, and anonymous architectural design, which involves both the factory's façade and the quality of indoor spaces (Armondi, 2013; Bennet, 1952; Busquets i Fabregas, 2007; Capra & Gulli, 2008; Cassatella & Gambino, 2013; Cipriani, 2012; Tandy, 1975). The typical image is of massive grey containers, sided by a succession of unpleasant auxiliart space and structures, such as service areas, settling basins, and raw material silos.

The careless appearance of most sites and the frequent neglect of their maintenance not only contributes to a negative image of the company among workers, visitors, local residents, and consumers, but can also accelerate the deterioration of structures and technological components because of the actions of aggressive pollutants on the building's surface (Busquets i Fabregas, 2007; O.P.U.Q.P., 2015). The same lack of design quality applies to indoor spaces, where no attention is usually given to colour, dimension, or materials, but also to the availability of natural light and pleasant outward views (Boeri et al., 2013).

**Water resource management**. Water use by industry at site-level deals with both the large volume of the resource required and its degree of contamination when leaving the factory. In Europe, total water use by manufacturing varies greatly, but the industrial sector is one of the major consumers along with households (Eurostat, 2020b): freshwater use by NACE C[5] was on average about 30 $m^3$ per inhabitant in 2018, varying from 193 (Netherlands) to 4 (Cyprus). The vast majority of this water is not treated after processing (Bortolini et al., 2017) and contributes to the pollution of water ways and groundwater. In this regard, major water contaminants accounted to industry in 2017 in EU-33 were organic carbon, nitrogen (N), phosphorus (P), and toxic metals (Cd, Hg, Pb and Ni) (European Environmental Agency, 2019b).

However, water pollution due to stormwater run-off, and industrial waste streams do not necessarily introduce chemicals such as these into bodies of water. Much more frequently it occurs thermal pollution that reduces the amount of oxygen in the water and, as a consequence, compromises its ability to support life (Tandy, 1975).

**Other resource management and pollution**. Many other natural resources are involved beyond water. Manufacturing affects the balance of both local and global ecosystems by means of intensive production processes, obsolete building technical services, poor construction techniques, and inefficient logistics (Delpiano & Trombetti, 2011), for instance by using both renewable and non-renewable resources, significant amounts of energy, and the release of polluting emissions and harmful waste into terrestrial and marine environments (Duflou et al., 2012; Pal, 2002). Concerning atmospheric contaminants, in the vast majority of cases they derive

[5] NACE is the Statistical Classification of Economic Activities in the European Community, and the letter C refers to Manufacturing.

from fossil fuel combustions that supply energy and fuel machine operations (Kaur et al., 2016). In 2017 European industry accounted for, in decreasing order, Carbon dioxide emissions ($CO_2$), Non-methane volatile organic compounds (NMVOCs), Toxic metals (Cd, Hg and Pb), Nitrogen oxide (NOx), particular matter (PM10), and Sulphur oxides (SOx) (European Environmental Agency, 2019b). Specifically, despite decreasing by 35% from 1990, industry was still the third largest Greenhouse Gas (GHG) emitter in the EU in 2018, after energy supply and transport (European Environmental Agency, 2019a). Hence atmospheric emissions related to factories are certainly a major concern.

However, industrial waste may also contain substances that alter the natural functions of the soil and are frequently harmful to flora, fauna, and human health. For instance, heavy metals, mineral oil, aromatic and polycyclic aromatic hydrocarbons, chlorine, and cyanides were the most common contaminants likely to affect soil, sludge and sediment in 2011 in the EU-33 (European Environmental Agency, 2019b).

## Needs and Trends to Mitigate the Impact

For many decades, all these negative consequences have been ignored or outsourced in the name of economic development and profitability. The long-lasting economic crises, the dramatic acceleration of environmental disasters and climate change, as well as the growth of societal issues, means that full support by all stakeholders is required. The multifaceted negative impacts of factories on the landscape must be mitigated, or even turned into a positive impact. To this end, the vast and complex industrial asset must be retroffited, remediated, or rehabilitated by considering all the relevant interlacing issues. Industry is one of the main contributors to the global crisis, but luckily it also has a lot of room for improvement. Each company and entrepreneur should consider their role and begin to include in their near-term scenarios the environmental and social negative externalities,[6] and potentially turn them into benefits for all. At the same time, policymakers and designers can strongly support this change by their actions.

In response, the notion of sustainable manufacturing has progressively emerged over decades, addressing, at least theoretically, all the three principles of sustainability (i.e. environment, economy, and society). Topics in this global challenge range from energy and resource efficiency to waste reduction; from contributing to the local economy to investing in infrastructure; from fair working conditions to good community relations (OECD, 2011).

An entrepreneurial change of perspective has not been the nly driving force. Technological advancements have also contributed. And equally, policy support to this universal transition has been crucial. The first environmental regulations on industry

[6] An external cost, also known as a negative externality, occurs when the social and economic activities of one group of people have an influence on another group and this impact is not fully justified or compensated for by the first group (Delpiano & Trombetti, 2011).

date back to the 1980s, following important milestones including the reports *Limits to growth* (Meadows et al., 1972) and *Our Common Future* (The Brundtland Commission, 1987). In the following decades, several strategies, programmes, and regulations have been introduced to push the manufacturing sector further in adopting more sustainable approaches. This history will be described further in the following chapter.

Today, the world 2030 Agenda for Sustainable Development has set many goals that target industry in order to protect the earth from degradation and climate change (United Nations, 2015). In particular, Sustainable Development Goals (SDGs) n.9 Industry, innovation and infrastructure, n.11 Sustainable cities and communities, and n.12 Responsible consumption and production frame political efforts worldwide.

At the European level, new policies and regulations are developing under the wide and general umbrella of the European Green Deal and the New European Bauhaus. On the one hand, the EU Green Deal (COM(2019) 640 Final, 2019) established the Union's commitment to tackle climate change and environmental challenges, directly engaging industry in several key goals. Specifically: by mobilising industry in order to achieve a clean and circular economy, and building and renovating in an energy and resource-efficient way. The Commission also stresses the need to support consumer choice through information retrieved by reliable standardised methodologies to prove the 'green claims' of firms. From this derives the new industrial strategy that addresses the twin challenge of a green and digital transformation of manufacturing (A New Industrial Strategy for Europe. COM(2020) 102 Final, 2020).

Meanwhile the New European Bauhaus (NEB) aims at bringing "the EU Green Deal to life in an attractive, and innovative and human-centred way". To this end, it envisions the future of the built environment based on three key pillars: beautiful, sustainable, and together (European Union, 2020). Sustainable buildings should go beyond being merely energy and resource efficient by also including a qualitative, aesthetic and human dimension. Thus, though it is not specific for factories, NEB emphasises the relevance of perceptual-aesthetic and socio-cultural aspects of the landscape in addition to the environmental ones.

In this framework, penalties for lack of compliance, tax benefits, and economic incentives have been implemented worldwide to lessen the environmental impact of manufacturing (Despeisse, 2013). In particular, great effort has been spent in the attention given to resource efficiency and the reduction of polluting emissions (May et al., 2016), but it is the industrial process that has been the main area of focus. Even if only in a small way, some guidelines and regulations have also addressed the sustainable conversion of existing facilities and the good design of new constructions. Some exceptional experiences led to the design of the so-called *Environmental fair factories* (Dombrowski & Riechel, 2013) and some others even theorised the *Positive impact factory* (Herrmann et al., 2015), which enrich the landscape instead of detracting from it.

Gradually, some planning, architectural, and social aspects have also been addressed in a few guidelines and studies (Boeri et al., 2013; O.P.U.Q.P., 2015; Tandy, 1975). However, in spite of the emerging body of literature proving that the perceptual-aesthetic harmonisation of buildings into the landscape is as important as any environmental and social considerations, this is still given little attention within industry. The reason for this is most likely that different aspects of a landscape are not considered together, and thus perceptual and social qualities of factories are largely seen as additional costs rather than benefits.

Instead, the triple bottom line of sustainable development must be addressed as a whole in order to get sustainable factories and mitigate their effects in daily practices. This requires concerted effort to reduce local and global pressures on the ecosystem, increase the value of the territory, and protect workers and local communities from industry-related disruptions. At the same time, beyond being environmentally and socially friendly, factories must be economically sound (Mueller et al., 2014). Several insights in both the scientific literature and examples from the real world suggest that this is possible if a holistic approach to sustainability is adopted.

Now is the time for those seeds of change to sprout, by boosting companies' economics while keeping sustainability at the forefront in the design and implementation of their facilities. Those who adapt and act now will gain the most, as the entire world—from policymakers to consumers—is increasingly supporting organisations that help to achieve sustainable development goals.

This book will address both the retrofitting of existing factories and the construction of new sites, but in particular it will underline the importance of *mitigating* the negative effects of what is already in place, as it represents the most pressing and difficult challenge we are currently facing, as companies, as architects, as policymakers, as neighbours, as consumers. Therefore, "Designing sustainable factories" means to conceive, construct, and operate industrial sites that are respectful of the physical environment, compatible with aesthetic-perceptual features of the place, and offers a healthy and stimulating working and living environment to all. In other words, factories that have little if any negative impact on the landscape or, even better, that have a positive impact.

## References

A New Industrial Strategy for Europe. COM(2020) 102 final, (2020). https://ec.europa.eu/info/sites/info/files/communication-eu-industrial-strategy-march-2020_en.pdf.

A SME Strategy for a sustainable and digital Europe. COM(2020) 103 final, (2020). https://eur-lex.europa.eu/legal-content/EN/TXT/PDF/?uri=CELEX:52020DC0103&from=EN.

Allegri, D. (2015). Modelli e tecnologie per la valorizzazione di paesaggi complessi. *TECHNE, 10*, 93–101. https://doi.org/10.13128/Techne-17505.

Armondi, S. (2013). Fare paesaggio con gli insediamenti produttivi. Riti e ritmi. In *XXXIV Conferenza Italiana Di Scienze Regionali* (pp. 1–18).

Ball, P. D., Despeisse, M., Evans, S., Greenough, R. M., Hope, S. B., & Kerrigan, R., et al.. (2013). Factory modelling: Combining energy modelling for buildings and production systems. *IFIP Advances in Information and Communication Technology, 397*(PART 1), 158–165. https://doi.org/10.1007/978-3-642-40352-1_21.

Bennet, W. R. (1952). Amenities in factory design. *Journal of the Royal Society of Arts, 100*(4878), 619–634. https://www.jstor.org/stable/41365418.

Boeri, C., Braz de Oliveira, I., & Giambruno, M. C. (2013). Colour design and industrial landscape. In *III CITCEM Conference*.

Bortolini, M., Gamberi, M., Pilati, F., Regattieri, A., & Accorsi, R. (2017). Design of an innovative plant for the wastewater recovery and purification in the food & beverage industry. In G. Campana (Ed.), *Sustainable design and manufacturing 2017* (vol. 68, pp. 691–700). Berlin: Springer. https://doi.org/10.1007/978-3-319-57078-5.

Busquets i Fabregas, J. (Ed.). (2007). *Per una corretta visione del paesaggio: linee guida* (Traduzione). Generalitat de Catalunya. http://www.paysmed.net/.

Capra, G. L., & Gulli, L. (2008). Nuovi modelli per gli insediamenti produttivi. In Raccolta degli atti dei convegni organizzati dall'Agenzia per la Qualità Urbana e Architettonica (Ed.), *Parma bellezza capitale. Proposte di qualità urbana* (pp. 83–106). http://servizioqualitaurbana.comune.parma.it/qualita_urbana/webpage.asp?ID=6&page=1&direct=true&IdMenu=19.

Cassatella, C., & Gambino, R. (2013). Linee guida per i paesaggi industriali, un'esperienza di ricerca. In *XXXIV Conferenza Italiana Di Scienze Regionali* (pp. 1–14).

Ciaramella, A., & Puglisi, V. (2016). Le caratteristiche degli immobili industriali. *Il Nuovo Cantiere, 2*, 82–84. http://www.ilnuovocantiere.it/le-caratteristiche-degli-immobili-industriali/.

Cipriani, L. (2012). *Vita e morte del paesaggio industriale. Indagini e proposte per la marca trevigiana in trasformazione*. Università degli Studi di Trento.

Delpiano, A., & Trombetti, G. (Ed.), (2011). *EnergiApea. Vincere la sfida energetica* (RicercaeProgetto—Galassi Mingozzi e associati). Tipografia metropolitana Bologna.

Despeisse, M. (2013). *Sustainable manufacturing tactics and improvement methodology. A structured and systematic approach to identify improvement opportunities*. School of Applied Science, Cranfield University. https://doi.org/10.1017/CBO9781107415324.004.

Dombrowski, U., & Riechel, C. (2013). Sustainable factory profile: A concept to support the design of future sustainable industries. In *11th Global Conference on Sustainable Manufacturing* (pp. 72–77).

Duflou, J. R., Sutherland, J. W., Dornfeld, D., Herrmann, C., Jeswiet, J., Kara, S., Hauschild, M., & Kellens, K. (2012). Towards energy and resource efficient manufacturing: A processes and systems approach. *CIRP Annals—Manufacturing Technology, 61*, 587–609. https://doi.org/10.1016/j.cirp.2012.05.002.

European Landscape Convention, (2000). http://www.coe.int/en/web/conventions/full-list/-/conventions/treaty/176?_coeconventions_WAR_coeconventionsportlet_languageId=it_IT.

European Environmental Agency. (2019a). *Greenhouse gas emissions by aggregated sector*. https://www.eea.europa.eu/ds_resolveuid/a57bad88e0824038ba6a7aabdc06a63b.

European Environmental Agency. (2019b). *Industrial pollution in Europe*. https://www.eea.europa.eu/data-and-maps/indicators/industrial-pollution-in-europe-3/assessment.

European Union. (2020). *New European bauhaus*. https://europa.eu/new-european-bauhaus/about/about-initiative_en.

Eurostat. (2008). *Rev. 2–statistical classification of economic activities in the European community*. https://ec.europa.eu/eurostat/ramon/nomenclatures/index.cfm?TargetUrl=LST_NOM_DTL&StrNom=NACE_REV2&StrLanguageCode=EN&IntPcKey=18496334&StrLayoutCode=HIERARCHIC&IntCurrentPage=1.

Eurostat. (2020a). *Manufacturing statistics - NACE Rev. 2*. https://ec.europa.eu/eurostat/statistics-explained/index.php?title=Manufacturing_statistics_-_NACE_Rev._2#:~:text=Themanufacturingsectoremployedmore,intheEUin2017.&text=Almost2millionenterpriseswere,intheEUin2017.

Eurostat. (2020b). *Water Statistics*. https://ec.europa.eu/eurostat/statistics-explained/index.php?title=Water_statistics#Water_uses.

Gandolfi, C. (Ed.), (2011). *Paesaggio e Attrattività*.

Harari, Y. N. (2011). *From animals into gods: A brief history of humankind. (Sapiens. Da animali a dèi: Breve storia dell'umanità)* (G. Bernardi (Ed.); Trad. ita.). Bompiani.

Herrmann, C., Blume, S., Kurle, D., Schmidt, C., & Thiede, S. (2015). The positive impact factory—transition from eco-efficiency to eco-effectiveness strategies in manufacturing. *Procedia CIRP, 29*, 19–27. https://doi.org/10.1016/j.procir.2015.02.066.
Kaur, G. P., Gupta, P., & Syal, M. (2016). Adoption of green practices in industrial buildings: An action research on capacity building of stakeholders towards green factories. *International Journal of Sustainable Land Use and Urban Planning, 3*(2), 1–12.
Lombardia, R. (2011). *Paesaggi periurbani. Linee guida paesaggistiche per il governo del territorio / Periurban landscapes. Landscape planning guidelines*. http://territorio.regione.emilia-romagna.it/paesaggio/pubblicazioni/paesaggi-periurbani.
May, G., Stahl, B., & Taisch, M. (2016). Energy management in manufacturing: Toward eco-factories of the future—a focus group study. *Applied Energy, 164*, 628–638. https://doi.org/10.1016/j.apenergy.2015.11.044.
Meadows, D. H., Meadows, D. L., Randers, J., & Behrens, W. W. (1972). *The limits to growth: A report for the club of Romes project on the predicament of mankind*. Universe Books.
Mueller, F., Cannata, A., & Herrmann, C. (2014). Procedure of modular green factory planning to enhance collaboration and decision making. *Procedia CIRP, 15*, 331–336. https://doi.org/10.1016/j.procir.2014.06.071.
O.P.U.Q.P. (2015). *Linee guida per i paesaggi industriali in Sardegna*.
OECD. (2011). *OECD sustainable manufacturing tookit—start-up guide*. www.oecd.org/innovation/green/toolkit.
Pal, U. (2002). Identifying the path to successful green manufacturing. *Journal of Minerals, Metals and Materials Society, 54*(5), 25. https://doi.org/10.1007/BF02701691.
Provincia di Bologna—Servizio Artigianato Commercio e Industria. (2009). *Risparmio energetico negli edifici ad uso produttivo* (pp. 1–122).
Provincia di Milano. (2012). *Adeguamento del PTCP alla LR 12/2005. Repertorio delle misure di mitigazione e compensazione paesistico-ambientali* (pp. 1–193). Piano Territoriale di Coordinamento Provinciale.
Reichardt, J. (2015). Site planning from space view. In H.-P. Wiendahl, J. Reichardt, & P. Nyhuis (Eds.), *Handbook factory planning and design* (pp. 331–339). Springer.
Settis, S. (2010). *Paesaggio Costituzione Cemento*. Einaudi.
Tandy, C. (1975). *Landscape of industry*. Leonard Hill Books.
The European Green Deal. COM(2019) 640 final, (2019). https://eur-lex.europa.eu/legal-content/EN/TXT/?uri=CELEX:52019DC0640.
The Brundtland Commission. (1987). *Our common future*. The Report of the World Commission on Environment and Development.
Tolli, M., Recanatesi, F., Piccinno, M., & Leone, A. (2016). The assessment of aesthetic and perceptual aspects within environmental impact assessment of renewable energy projects in Italy. *Environmental Impact Assessment Review, 57*, 10–17. https://doi.org/10.1016/j.eiar.2015.10.005.
Tudor, C. (2014). An approach to landscape character assessment. In *Natural England* (Issue October). https://assets.publishing.service.gov.uk/government/uploads/system/uploads/attachment_data/file/691184/landscape-character-assessment.pdf.
Turri, E. (1979). *Semiologia del paesaggio italiano*. Longanesi.
U.S. GBC. (2014). *LEED core conceptsg guide*. U.S. Green Building Council.
United Nations. (2015). *Transforming our world: The 2030 agenda for sustainable development (A/RES/70/1)*. https://documents-dds-ny.un.org/doc/UNDOC/GEN/N15/291/89/PDF/N1529189.pdf?OpenElement.
Veglia, A., & Bono, S. (2007). Progettazione e valutazione di uno stabilimento industriale. In Celid (Ed.), *Bioedilizia Italia - Atti del Congresso Nazionale di Architettura Sostenibile* (pp. 1–8). http://www.patdesign.it/wp-content/uploads/2014/03/PAPER_industria.pdf.

# Chapter 2
# Level of Awareness and Economic Constraints as Barriers to Sustainable Factories

**Abstract** Industrial sites have always interfered with the landscape, but the extent and the nature of the impact have changed considerably over the centuries. So, what has driven the design of factories in the past? Why has their sustainability been hampered? As general awareness on the topic has evolved, influence on the impacts of industrial sites has evolved with it. The same holds true for economic aspects. Today, both industry and policymakers have largely recognised the reduction of environmental impacts as crucial to sustainable development. Perceptual-aesthetic and social issues related to factories have also become increasingly important for the protection of the landscape and its local community. Considerable effort has been spent on reducing the environmental effects of manufacturing at the process level, not only because this stage accounts for the majority of resource consumption and emissions, but also because the economic benefits are more easily understood by firms. Less attention has been paid to the performance of facilities at site and building levels, where sustainability is still largely perceived as simply an additional cost. Moreover, while big companies find it easier to invest resources in design and to communicate their sustainability, small and medium-sized businesses, which account for most European firms, struggle with resource shortages and a lack of expertise that prevents them from undertaking the change.

**Keywords** Industrial cities · Profit · Green factories · Environmental concern · Cost of sustainability

The 2001 Nobel prize laureate, Joseph Stiglitz, in his *Making globalization work*, claimed that sooner or later the world would have to put itself in the frame of mind for making certain changes, and that the question is not so much whether those changes will occur, but when they will occur, and—more importantly—if they will come before or after another series of global disasters (Stiglitz, 2007). The industrial sector is particularly affected by this as it is a big player in the global-changing pathway. The need to change actual models of production has been acknowledged worldwide. Now the question is, rather, how can this change be effectively supported and boosted?

In particular, when considering industrial architecture, prior to any discussion concerning the design of sustainable factories, it is crucial to understand the historical and present-day barriers that prevent companies from respecting the environment,

 
L. Marchi, *Designing Sustainable Factories*, Advances in Global Change Research 72,
https://doi.org/10.1007/978-3-030-92227-6_2

the scenery, and the local community through the design and implementation of their facilities. In other words, what factors have guided and continue to guide the design of industrial sites. Indeed, only after identifying the obstacles that must be overcome can we act effectively in order to make this change happen.

## A Brief Historical Outline

Regardless of the period or the form it has assumed, industry has always had an impact on the landscape (Tandy, 1975). Although we are used to thinking of the birth of factories as an expression of the Industrial Revolution, in fact, man has always sought ways and means to exploit natural resources to produce artifacts. Accordingly, through history, spaces of production have been primarily designed in order to best serve function and profit. The impact on the landscape has gradually grown in scope and intensity. Although public awareness has evolved over time and influenced the impact itself, economic constraints have largely hindered attempts to integrate the factory into the landscape. In the name of progress, the triple bottom line of sustainability—people, planet, and profit—has largely been ignored.

However, not always have factories been such a blight on the landscape as they are today. The extent and the nature of their interference has changed considerably over the centuries. Placement dynamics, technological advancement, and architectural features are among the most relevant factors that have historically shaped their impact, each factor being closely interlinked with economic considerations too.

**Placement**. Many variables have affected the growth and setting of industry, but financial convenience—in many forms—has largely prevailed.

In the early times, the impact of proto-industries on the landscapes was not so severe since they were intimately linked to their place of production, their features and location being crucial for the process of production itself. Proximity to raw material extraction or energy sources had been preferential criteria for many years. The first "industrial apparatus" were located close to waterways for the energy and important raw materials they could provide for industrial processes (Vitali, 2014). Likewise, the perishable nature of food and the available means of transportation gave a prevailing logic to the location of industry for many centuries: wineries were close to vineyards, granaries to wheat fields, slaughterhouses to farms, bakeries to villages (Clark, 2005; Murthy, 2016). The simple necessities of these early times made the overall impact of industry limited: respecting the environment and its natural rhythm appeared to be essential for both living and producing goods.

Manufacturing had gradually grown during the Middle Ages, but proximity to resource extraction and energy sources remained critical, and the presence of large sites in the landscape remained limited for a long time (Ashton, 1948).

The relationship between site and production changed drastically during the Industrial Revolution (seventeenth century), when new energy sources and means of transportation allowed for longer distances at a lower cost (Harari, 2011). At this point, progress and profit-making started to hold sway over all other aspects. The rapid rise of manufacturing resulted in a high concentration of factories in cities and suburbs. Economic and environmental balances were decoupled, with the resulting implications on the ecosystem being widely undervalued by businesses and organisations. However, because major hygienic problems related to factories began to emerge, public awareness of industry's impact on the living environment started to grow. Consequently, many factories were relocated out of the city to the countryside in order to avoid health laws and compliancy costs. Unfortunately, on the other hand, constructing on virgin areas "legitimated" the design of facilities without concern for their surroundings, as they were now out of public view.

In the nineteenth century, the poor conditions of workers in cities reached significant peaks, provoking rebellions. People became aware of the severe social and health consequences of factories, and the demand for separate living and working spaces grew exponentially (Engels, 1845). Often initiated on the idealistic momentum of *garden cities* (Howard, 1902), Saltaire and Port Sunlight in the United Kingdom, and Crespi d'Adda and Schio in Italy[1] are some clear examples. These are integrated projects in which a village develops around a factory, in close relationship with it, placing the health and well-being of workers at the heart of planning processes. They remained excellent but isolated experiments.

The twentieth century received these ideals only in part. Principally, it exacerbated an already critical situation. Some partially effective placement criteria dealt with worker proximity, first, and cluster organisation for sharing services and infrastructure, later (Jiang et al., 2017). This latter concern, in particular, was aimed at seeking economic and environmental sustainability. Although zoning was intended to organise and limit urban growth, in many cases it fed the construction of "islands," where space for social services, leisure, and beauty was considered unnecessary. Meanwhile, ineffective planning rules led to chaotic additions and new constructions scattered throughout the territory, leading to the further congestion of cities and suburbs or affecting the balance of the countryside.

**Technological level**. Similarly throughout history different tools, techniques, and energy sources employed by various forms of production have required progressively more extensive adaptations of the natural surroundings in order to increase the effectiveness of machinesdevices and processes. As the pressure on the environment increased, so did public consciousness, but for most companies intention has not resulted in action..

To begin with, activities and practices (mainly man and animal work-based) did not produce serious negative effects on the ecosystem or the scenery. But soon dramatic landscape transformations took place. In Greece, serious occurrences of deforestation are noted in the literature since the sixteenth century BC (Bromehead, 1954). The exploitation of the environment and the scenery detraction intensified in

---

[1] In this regard, Tandy (1975) reports some examples such as Saltaire (1855–75) and Port Sunlight (1888–1920+). Beyond the importance of Port Sunlight as an experiment, it is worth mentioning that its founder, William Lever (inventor of the laundry soap in bulk), is also considered one of the father of urbanism, having founded a planning school in Liverpool in 1909, the School of Architecture's Department of Civic Engineer. In Italy, it values a citation the industrial village of Crespi d'Adda (1878), as an integrated form of residence and work, and the new working-class district of Schio (1872), where Alessandro Rossi built an "ideal city" for its textile employees.

the Roman era and beyond, principally due to the widespread construction of road and water infrastructure. Progressively, in the Middle Ages, mechanical advancement and manufacturing development increased the exploitation of natural resources, thus detrimental effects on the environment arose. In the seventeenth century, water mills, followed by windmills, began to create a landscape of production but did not yet drastically harm the balance of the ecosystem.

The invention of the steam engine, which replaced animal traction and human strength, profoundly altered the relationship between the production site and the landscape. The ancient rules of environmental respect, attachment to place, identity, and local culture were severely compromised with the advent of the industrial society that freely shaped the natural landscape for its convenience (Turri, 1998). The Industrial Revolution (eighteenth century) drastically affected the balance between production and the landscape. The long line of technological discoveries and innovations, first driven by mechanics and chemistry, and then by informatics, transformed industrial processes and, as a result, the features of the facilities. The impact on the landscape became increasingly intense, both in terms of natural resource depletion and pollution and the size of factories. Factories were simply considered places for producing goods, not a part of people's living environment.

In the middle of the nineteenth century, the Second Industrial Revolution made the growth exponential. On the one hand, exceptional technological breakthroughs offered a solid stimulus to manufacturing. On the other hand, the negative effects on the environment became stronger and more widespread.

Along with the escalation of industry-caused environmental and social problems, public awareness on the issue increased. The environmental and economic crises of the 1970s represents a milestone on this path. It promoted the idea that no industrialist could conceive of a new plant—or an existing factory's expansion—without considering the environmental implications as well as the potential repercussions on public opinion. From that moment on, the interest of companies towards sustainability would be constantly reinforced. Nevertheless, the general low quality and high impact of industries that characterises our everyday landscapes suggest that there is a significant gap between good intentions and what is really put into action (i.e., intention-action gap).

Technological advancements have become so extensive that it is hard to generalise their environmental and social implications from the second half of the twentieth century onwards. Rather, it is interesting to focus on the awareness of policymakers and companies themselves and their approach to the mitigation of any impact that relates to technological innovation.

According to Azapagic and Perdan (2000), between the early-1970s and the mid-1980s the firms' general behaviour can be defined as "reactive". Globally, this phase refers to an early awareness of the effects that industry has on the natural dimension of the landscape. Therefore, during this time, the first steps were taken to reduce the environmental effect of existing manufacturing plants, mainly in response to new constraints and regulatory limits, as well as the ongoing energy crisis. Despite the need to dramatically transform industrial processes by balancing production requirements with its repercussions on social and environmental levels, the measures in place were still modest, owing to the extreme complexity of the issue (CAETS, 1996). Most interventions were end-of-pipe, which means they were solutions to minimise the pollution released into the environment once it had already been generated. As much as these technologies were low-cost and easy to implement, their effectiveness was limited.

Then came the "pro-active" phase. Up to the early 1990s institutions and customers demanded more and more sustainability from businesses. The approach used thus far would be unsustainable in the long run, and so it became clear tha any intervention had to occur at the root of the problem. Costs rose to meet these demands. At the same time, several concepts and approaches linked with 'industrial ecology' emerged (Grant, 1997), highlighting possible economic benefits of the operation. Many companies realised that greener manufacturing, by reducing resource waste and boosting process efficiency, not only improved their environmental profile, but could also cut operational costs and increase profits. Aside from improving production processes to save energy and resources, in this stage some strategies addressed the architectural consistency of manufacturing facilities, maximizing their efficiency to decrease operating costs. Despite being supported by academics, the implementation of mitigation strategies was mainly restricted to big corporations with greater spending capacities.

Later on, since the 1990s, environmental sustainability has become a core value for many firms, not only because of the economic payback, but also because of the strengthening of the corporation's image in the eyes of consumers and local communities. This is the phase when "environmental certifications and reports" gained traction. They quickly spread as procedures to improve performance as well as to promote and communicate the company's profile. According to a survey by the *KPMG International Environment Network*, between 1993 and 1999 the percentage of companies globally producing environmental reports grew from 13 to 24% (Punte & Iansen-Rogers, 1999): the most common references were the ISO 14001[2] standard and EMAS[3] voluntary certification system. Steered by improving the energy efficiency of industrial sites, the first 'carbon neutral factories' or 'environmental fair factories' were established. Unfortunately, these remained isolated cases rather than viable models to be applied on a large scale (Dombrowski & Riechel, 2013).

Meanwhile, the early experiences of ecological industrial areas gained attention worldwide: from the Eco-industrial parks (Lowe et al., 1996), to the Ecologically Equipped Productive Areas (APEA, in Italy). Regardless of the specific names, they are communities of manufacturing and service businesses that insist on the same area and collaborate to improve their environmental, economic, and social performances,

[2] ISO 14001 is a set of environmental management standards designed to help organizations in reducing the negative impact of their operations on the environment; complying with applicable laws, regulations, and other environmentally oriented requirements; and continuously improving in the above areas. They set requirements for environmental reporting, but they do not establish thresholds.

[3] Eco-Management and Audit Scheme (EMAS) is the European system to environmental management and audit to which companies can voluntarily adhere, to give information on their environmental performances and obtain.

aiming at achieving collective benefits greater than the sum of their parts (Provincia di Bologna, 2008). Once again, many of these good practices and approaches have remained the exception to a widespread profit-driven attitude.

**Architectural features.** Across the centuries, the progressive enlargment of production sites in response to economic growth and the strengthening of profit-driven attitudes has coincided in most cases with a deterioration of the architectural quality of the site. Production sites have traditionally been dedicated to work, and little attention has been paid to the negative effects that their low-quality design has had on workers and locals. Beauty has largely been regarded as needless in manufacturing, so blight and ugliness seem inevitable by-products of businesses (Maguire et al., 1997).

In the beginning, the visual impact of farms, mills, and even quarries was limited. In general, neither their small dimension nor their aesthetic features (e.g., built with local materials and techniques) produced conflicting effects on the landscape. What remains of that time is mostly deemed romantic and poetic nowadays.

Then, as craftsmen gathered in groups in the Middle Ages, production sites grew in size and assumed new forms. During the Renaissance, in the fourteenth century, the concept of a 'factory' took shape (Tandy, 1975): places of production could be clearly distinguished by the shape and character of the houses and workshops.

By the seventeenth century, large industrial sites were scattered throughout the landscape, both in urban and rural settings. Their shape and features varied according to their function: the landscape became filled with high furnaces and smoking chimneys; mills grew up to eight floors, becoming visible from long distances; the owners of large factories built villages for their workers, complete with schools and services; the deep extraction of coal devastated rural landscapes while leaving mountains of debris; water channels became widespread, but in most cases their design kept in harmony with the landscape. It is clear that the different functions and dimensions of each site generated different positive and/or negative impacts on the landscape.

The nineteenth century saw a radical shift from the domestic character of the house-shop to the development of specialised production structures, as well as the transformation of vast production-commercial complexes into industrial ghettos. This gradually highlighted the social detriments of manufacturing and the general awareness of the population increased. In the twentieth century, the dichotomy among small family-run businesses and large industrial areas persisted, but both forms—and everything in the middle—were associated with a general progressive decline in the relations of the production site with the context.

Apart from a few notable examples in which landscape and architectural quality represented central design criteria, industrial structures became increasingly focused on trivial and repetitive solutions as a mass response to manufacturing needs. Factories were configured only as simple containers to protect machinery. *Forms follow function* and nothing else.

Meanwhile, a small but notable group of companies took the first step towards a new corporate architecture. From Peter Behrens' AEG Turbine Factory, in Germany (1909) to Gino Valle's Fantoni Factory, in Italy (1973), these were facilities intended to communicate the firm's values and express progress. These are just two clear examples of this approach, which provided important insights into the future of factory design but, on the other hand, focused once again on just one aspect: the aesthetic, or, more precisely, the expression of the machine aesthetic.

Coming back to the leading functional approach, many public authorities introduced standards and guidelines for production sites between the 1960s and the 1970s, mainly to cope with the growing environmental problems (Grant, 1997). These affected the visual quality of factories as well, sometimes positively, other times negatively. Some defined the minimum distance between buildings, others defined measures and treatments for green spaces or aesthetic requirements for the façades. Meanwhile, increasingly adopted energy-saving strategies resulted in a negative impact on the architectural consistency of the factory. Windows were avoided as dispersant surfaces, in addition, indoor ventilation and the availability of daylight were considerably curtailed in order to limit dispersions. This affected the visual quality of facilities, but also the psycho-physical health of workers. In other words, instead of integrating, environmentally friendly features, they were often implemented at the expense of aesthetic and visual comfort. Thus, the replication of identical industrial warehouses were mostly the result of cost-effectiveness and functional efficiency criteria, or a flat manual application by selecting prefabricated components from catalogues (Veglia & Bono, 2007). In some cases the design of factories even resulted from functional optimisation through mathematical models (Ciaramella & Celani, 2019).

Luckily, some interest in the topic began to grow in line with the advent of concepts such as "visual pollution"[4] (Maguire et al., 1997), on the one hand, and "brandscape"[5] (Klingmann, 2007) on the other. This latter concept, however, entails a high risk of superficial design. Industrial warehouses and plants gradually took on the characteristics of concrete grey structures, where the symbol, rather than the form, dominated the architecture. The building's envelope became a sign, and the façade became a surface to be planned according to precise motion-driven measures to control the kinetic experience of the spaces from the perspective of a car window, when enough resources were available and the entrepreneur was sufficiently aware of the power of marketing. This is how Cipriani (2012) describes the industrial landscape of Northern Italy, where the image of the factory has been progressively reduced to mere advertisement (Capra & Gulli, 2008; Qualizza, 2012). In most situations, it translates to simply placing the firm's logo on the building, which is far from establishing any kind of relationship with the landscape (either socially or visually).

[4] Visual pollution is an aesthetic issue that refers to the effects of a place visual features on one's ability to enjoy a pleasant view. Billboards, storage of trash, antennas, electric wires, buildings, and car/traffic jam are often considered visual pollution.

[5] Brandscape is a neologism that combine brand (i.e. identity sign of a firm) and landscape. Anne Klingmann in her *Brandscape* argues that "buildings are not about where we work and live but who we imagine ourselves to be". The concept brings either the willingness of a company to strengthen its identity and marketing promotion by architecture, product, advertisements, either the social and cultural opportunities that emerge for the territorial attractiveness.

## Current Approaches and Existing Barriers

The brief historical outline shows that in recent times policymakers have encouraged the raising of awareness of the environmental, social, and aesthetic-perceptual impacts of factories on the landscape. On the other hand, a more conscious approach to factory design resulted from a general increase in the awareness of companies awareness and a growing role of consumers and citizens in influencing manufacturing.

In fact, there is evidence that firms themselves now recognise their own impacts more clearly and are trying to mitigate the effects (Ball & Jolly, 2015). In particular, great effort has been expended to lessen the environmental impact of manufacturing operations at process level, focusing on resource efficiency and the reduction of polluting emissions (May et al., 2016). Since on average industrial processes account for the greatest share of a factory's impact on the ecosystem, their increased efficiency can both reduce pressure on nature (reduced demand for energy and resources, lower emissions of pollutants) and direct costs. Instead, the importance of the environmental standard of a facility at site and building levels is still under-estimated.

Guidelines and incentives have been promoted with the aim of designing new efficient and environmentally friendly industrial sites, but the unsustainable level of our consumption models is based on the industrial sites that exist today (Delpiano & Trombetti, 2011). Indeed, the renovation or retrofitting of existing operating facilities are very often neglected by these measures. Even when considered, minimal interventions are often proposed and usually focus on the principle of efficiency alone. It should not only be new projects that are driven by a sustainable design approach; the retrofitting, renewal, and enhancement of existing factories should follow this approach too.

Social issues have also been recognised as important by public opinion and entrepreneurs. The key addressed concern is worker health and well-being, which is tackled through space quality and ergonomics, services, and indoor comfort. Environmental sustainability and social responsibility reports are spreading to support the communication of companies' performances. Within Corporate Social Responsibility[6] (CSR) particular importance is given to services and activities that support local communities or vulnerable areas in developing countries as social compensation. But entrepreneurs must pay attention as this conventional approach is increasingly expected to be authentic and not mere social washing, especially by younger generations (Rudominer, 2017). Like negative environmental externalities, it is no longer acceptable to compensate for social externalities via "surface charity": consumers are getting increasingly suspicious in this regard. Rather, a company's social initiatives could have greater value if their principle concern was the care of their employees.

In parallel, the importance of the perceptual-aesthetic quality of industrial facilities is gaining prominence, not only for boosting territorial attractiveness

[6] Defined as "corporate values and practices emphasizing employee and societal well-being" (Schönborn et al., 2019).

and protecting the community, but also for implicit company promotion through marketing. However, there is a high risk of incurring accusations of green-washing or beautification strategies, when green claims and visualisation are not substantiated by actual practices (Qualizza, 2012; Tandy, 1975). The greater the consumer interest and demand for sustainability, the greater the risk of companies presenting something as green that is just green-painted or achieved by simply installing wind turbines (Addy, 2008).

All in all, the level of awareness on the topic seems to be reasonable, and some of the economic benefits of sustainability have gradually become understood by businesses themselves. Consequently, it would be interesting to examine what is currently preventing companies from designing 360° sustainable facilities.

While several large corporations have already taken significant steps towards becoming more sustainable, small and medium-sized businesses—representing the vast majority of industries—still have difficulties related to costs, available resources, and lack of expertise (Kaur et al., 2016; OECD, 2011).

Unsurprisingly, the first and most important barrier is represented by the amount of money that is required. The cost of transformation is, according to Pojasek (2008), the only reason for a company not going green: some even see this as a matter of choice between firing employees and being environmentally responsible.

In Europe, for example, SMEs account for nearly the totality of businesses (more than 99%). Interestingly, their total gross value was only 38.5% in 2018, with the remaining 61.5% coming from large firms representing less than 1% of all businesses (Eurostat, 2020). This suggests that SMEs, which own and run thousands of small to medium-sized factories, have a far lower investment potential than larger corporations. So, it is understandable that their first goal is to survive and address any impending debts over retrofitting interventions which would require, in general, high upfront costs with a medium-to-long payback period (Delpiano & Trombetti, 2011). Unfortunately, the Covid-19 pandemic has aggravated the situation even more. Many small businesses did not survive, and for many others, serious economic challenges may have occurred, raising the risk of overlooking the "additional" cost of sustainability. That is why, to avoid taking steps backwards on the path to sustainable development, the EU recovery plan *Next Generation Europe* has stressed on coupling pandemic recovery and sustainability with the slogan "make it strong, make it green" (*Next Generation Europe*, n.d.).

Beyond economic issues, an extensive review of barriers and drivers of eco-factories has found out that institutional (e.g. lack of regulations, incentives, coordination) and organisational (e.g. culture, value, attitude) factors recur, but also awareness-, behavioural-, competence-, information- and technology-related obstacles are also often registered (May et al., 2016).

As a matter of fact, tightly connected with resource shortage and the issue of cost is the misleading perception that sustainability should still be mainly regarded as an extra feature and thus an additional cost. This perception is perhaps true when sustainability is included later in the design, or even during the construction process, but it has been proven false when a sustainable approach is included in design's early stages. Since a lack of information is another important barrier, several guidelines, books, and studies have gradually spread to illustrate the benefit of sustainable manufacturing and green factories. Some factories from the real world have also supported this spread of information. As did the dissemination of sustainability standards and certification (particularly ISO 14001), something which encouraged entrepreneurs to take action to optimise their factory's consumption and lessen its overall impact (Dombrowski & Riechel, 2013).

Today, regarding sustainable manufacturing as a whole, there is a consistent body of literature but it is extremely disorganised. Duplication of concepts and efforts, as well as a lack of awareness of the global impact of local improvements (i.e. rebound effect), create obstacles for the adoption of sustainable practices (Despeisse, 2013). A holistic approach is still lacking, and many companies are unaware of their actual impact, as well as possible mitigation actions they could take and their associated benefits. As a result, it appears that suitable incentives and awareness raising would be helpful in overcoming these barriers, not only for firms, but for all stakeholders, including practitioners and contractors who may also make suggestions and help drive the design and construction process of factories to be greener.

## References

Addy, R. (2008). Green is the goal. In *FOOD manufacture, Ottobre* (pp. 45–46). https://www.foodmanufacture.co.uk/Article/2008/09/29/Green-is-the-goal.

Ashton, T. S. (1948). *The industrial revolution 1760–1875.* Oxford University Press.

Azapagic, A., & Perdan, S. (2000). Indicators of sustainable development for industry: A general framework. *Trans IChemE, 78*(4), 243–261. https://doi.org/10.1205/095758200530763.

Ball, P., & Jolly, M. (2015). *Sustainable manufacturing for the future. Investigating the current and future landscape across the food and drink industry in Great Britain* (Issue June).

Bromehead, C. N. (1954). Mining and quarrying, in the 17th century. In C. Singer, E. J. Holmyard, & A. R. Hall (Eds.), *A history of technology* (vol.1). Oxford University Press.

CAETS. (1996). *The role of technology in environmentally sustainable development.*

Capra, G. L., & Gulli, L. (2008). Nuovi modelli per gli insediamenti produttivi. In raccolta degli atti dei convegni organizzati dall'Agenzia per la Qualità Urbana e Architettonica (Ed.), *Parma bellezza capitale. Proposte di qualità urbana* (pp. 83–106). http://servizioqualitaurbana.comune.parma.it/qualita_urbana/webpage.asp?ID=6&page=1&direct=true&IdMenu=19.

Ciaramella, A., & Celani, A. (2019). Industria 4.0 e manifattura in città: uno sviluppo verticale possibile. *TECHNE, 17*, 133–142. https://doi.org/10.13128/Techne-24009.

Cipriani, L. (2012). *Vita e morte del paesaggio industriale. Indagini e proposte per la marca trevigiana in trasformazione.* Università degli Studi di Trento.

Clark, J. P. (2005). Food plant design. Food engineering. In G. Barbosa-Canovas (Ed.), *Enciclopedia of life support systems* (pp. 683–696). EOLSS Publisher/UNESCO.

Delpiano, A., & Trombetti, G. (Ed.), (2011). *EnergiApea. Vincere la sfida energetica* (RicercaeProgetto—Galassi Mingozzi e associati (ed.)). Tipografia metropolitana Bologna.

Despeisse, M. (2013). *Sustainable manufacturing tactics and improvement methodology. A structured and systematic approach to identify improvement opportunities.* School of Applied Science, Cranfield University. https://doi.org/10.1017/CBO9781107415324.004.

Dombrowski, U., & Riechel, C. (2013). Sustainable factory profile: a concept to support the design of future sustainable industries. In *11th Global Conference on Sustainable Manufacturing* (pp. 72–77).

Engels, F. (1845). *Die Lage der arbeitenden Klasse in England (La situazione della classe operaia in Inghilterra, in base a osservazioni dirette e fonti autentiche (2011))* (Trad. italiana Cervetto A. (ed.)). Edizioni Lotta Comunista.

Eurostat. (2020). *Manufacturing statistics—NACE Rev. 2.* https://ec.europa.eu/eurostat/statistics-explained/index.php?title=Manufacturing_statistics_-_NACE_Rev._2#:~:text=Themanufacturingsectoremployedmore,intheEUin2017.&text=Almost2millionenterpriseswere,intheEUin2017.

Grant, J. (1997). Planning and designing for eco-efficiency industrial landscapes. *Journal of Cleaner Production, 5*(1–2), 75–78.

Harari, Y. N. (2011). *From animals into gods: A brief history of humankind. (Sapiens. Da animali a dèi: Breve storia dell'umanità)* (G. Bernardi (ed.); Trad. ita.). Bompiani.

Howard, E. (1902). *Garden cities of tomorrow.* Swan Sonnenschein & Co.

Jiang, G., Ma, W., Dingyang, Z., Qinglei, Z., & Ruijuan, Z. (2017). Agglomeration or dispersion? Industrial land-use pattern and its impacts in rural areas from China's township and village enterprises perspective. *Journal of Cleaner Production, 159*, 207–219. https://doi.org/10.1016/j.jclepro.2017.04.152.

Kaur, G. P., Gupta, P., & Syal, M. (2016). Adoption of green practices in industrial buildings: An action research on capacity building of stakeholders towards green factories. *International Journal of Sustainable Land Use and Urban Planning, 3*(2), 1–12.

Klingmann, A. (2007). *Brandscapes. Architecture in the experience economy.* MIT Press.

Lowe, E. A., Moran, S. R., & Holmes, D. B. (1996). *Fieldbook for the development of eco-industrial parks* (Final Repo). Indigo Development.

Maguire, M., Foote, R., & Vespe, F. (1997). Beauty as well as bread. *Journal of the American Planning Association, 63*(3), 317–328. https://doi.org/10.1080/01944369708975925.

May, G., Stahl, B., & Taisch, M. (2016). Energy management in manufacturing: Toward eco-factories of the future—a focus group study. *Applied Energy, 164*, 628–638. https://doi.org/10.1016/j.apenergy.2015.11.044.

Murthy, S. R. (2016). *Designing sustainable industrial systems. Principles—patterns—practices.* Institute for Manufacturing, Dept. Engineering, University of Cambridge.

*Next Generation Europe.* (n.d.). https://europa.eu/next-generation-eu. Last accessed 29 July 2021.

OECD. (2011). *OECD sustainable manufacturing tookit—Start-up guide.* www.oecd.org/innovation/green/toolkit.

Pojasek, R. B. (2008). When is green manufacturing green? *Manufacturing Engineering, 141*(2), 18–19. www.sme.org/manufacturingengineering.

Provincia di Bologna. (2008). *APEA—Aree Produttive Ecologicamente Attrezzate. Linee guida* (M. Cavallo (Ed.)).

Punte, S., & Iansen-Rogers, J. (1999). *KPMG international survey of environmental reporting 1999.*

Qualizza, G. (2012). Inventare lo spazio di una green factory. In M. Cavallo, P. Degli Esposti, & K. Konstantinou (Eds.), *Green marketing per le aree industriali. Metodologie, strumenti e pratiche* (pp. 220–238). Franco Angeli.

Rudominer, R. (2017). *Corporate social responsibility matters: Ignore millennials at your peril.* https://www.huffpost.com/entry/corporate-social-responsi_9_b_9155670?guccounter=1&guce_referrer=aHR0cHM6Ly9teWZvb2Rzb2Z0d2FyZS5jb20v&guce_referrer_sig=AQAAAJxt7Ikv_vhrW86h7Lws_4S7clA537YqaxHfJ9vUQurYtWaknqsEuKNF9sll5DgPVtOc7Tu_uAgpDrTJi3ny5NgtFTM_IN3w2as6.

Schönborn, G., Berlin, C., Pinzone, M., Hanisch, C., Georgoulias, K., & Lanz, M. (2019). Why social sustainability counts: The impact of corporate social sustainability culture on financial success. *Sustainable Production and Consumption, 17*, 1–10. https://doi.org/10.1016/j.spc.2018.08.008.

Stiglitz, J. E. (2007). *La globalizzazione che funziona* (G. Einaudi (Ed.); Trad. ital).

Tandy, C. (1975). *Landscape of industry.* Leonard Hill Books.

Turri, E. (1998). *Il paesaggio come teatro: dal territorio vissuto al territorio rappresentato.* Marsilio.

Veglia, A., & Bono, S. (2007). Progettazione e valutazione di uno stabilimento industriale. In Celid (Ed.), *Bioedilizia Italia—Atti del Congresso Nazionale di Architettura Sostenibile* (pp. 1–8). http://www.patdesign.it/wp-content/uploads/2014/03/PAPER_industria.pdf.

Vitali, P. (2014). I luoghi della produzione nei territori della deindustrializzazione. Il caso Bergamo. In S. Marini & F. De Matteis (Eds.), *La città della postproduzione* (Edizioni N, pp. 61–71). https://doi.org/10.4458/1921.

# Chapter 3
# Potential Drivers of the Change

**Abstract** Academics and policymakers now argue that social and environmental externalities, which for many years had simply been excluded from the purview of businesses, must be carefully considered, if shortcomings affecting the company's competitiveness are to be avoided, and growth opportunities are not to be lost. So, what can drive this change? Although some exemplary sustainablefactories demonstrate advantages that outweigh mere profit making, economic benefits are certainly crucial in pushing this change. This chapter therefore begins with a description of the direct and indirect economic benefits, with a particular focus on those that stem from resource and energy savings, incentives, green marketing and territorial attractiveness. In addition to these benefits, this chapter will examine the enhanced competitiveness that derives from an improved corporate image and an increase in sustainability communications, both in terms of the visual and perceived quality of the facility, and also of the care of its occupants and the local community. Finally, this chapter will consider the entrepreneurs' own sense of responsibility, impact understanding and the awareness that might drive the enhancement of a business' sustainability performance. In this framework, practical support to companies can also facilitate change. To this end, some tools have been developed to help companies assess and mitigate their impact, or even to turn their weaknesses into opportunities. Green Building Rating Systems and visual impact assessment tools are among the most common. However, these are generally too sectorial, and often fail in achieving an integrated approach to the topic.

**Keywords** Corporate image · Energy saving · Green marketing · Impact assessment · Incentives

Important steps have been taken to mitigate the impact of factories on the landscape, at different scales and from different angles. Despite the growing interest of business, political support, and contribution from academia, a unified vision capable of effectively managing the multiple dimensions of this topic is lacking. The many intertwining issues are still largely addressed separately, and this might be one of the main reason sustainability is perceived as an additional cost rather than a source of green opportunities.

 
L. Marchi, *Designing Sustainable Factories*, Advances in Global Change Research 72,
https://doi.org/10.1007/978-3-030-92227-6_3

So far, the interferences that need to be mitigated have been discussed, as well as the obstacles that have traditionally hampered the design of sustainable factories. This chapter will discuss the tools capable of overcoming those barriers and triggering the change. It seems plausible to affirm that the same kinds of factors that have guided the conventional profit-driven approach to factory design are the first to be positively changed into drivers (i.e. the level of awareness and economic issues). In other words, these can be considered using Meadows' definition of the "leverage points" of a system (Meadows, 1999): those areas, within a complex system such an organization, an economic system, the human body, a city, an ecosystem, where slight modifications can cause great changes.

Companies must be firmly convinced that designing more sustainable factories is worthwhile. They need proof that the procedure is beneficial and feasible. Therefore, the many pragmatic queries they might have need to be answered with facts. Questions such as: what are the tangible and intangible benefits for those companies operating with respect for nature, the scenery, and the local communities, both at local and large scales? What are the positive benefits on the territory that can justify economic investments by public or private consortia, as well as local administrative support? What are the customers interests in this matter and how can this motivate companies to seek optimum compatibility with the surrounding landscape? This chapter will attempt to answer these questions.

## Economic Drivers

Unsurprisingly, the first and perhaps most powerful driver are economic considerations. As a matter of fact, sustainable factories must be economically sound, beyond being environmentally-friendly, where the latter is usually considered only as a beneficial side effect by businesses (Pal, 2002). However, economic drivers refer to a great variety of factors, not necessarily related to direct profit. They span from resource and money savings to regulatory compliance and incentives, from increased marketing value to territorial attractiveness. In fact, both the direct and indirect economic benefits of sustainability can encourage entrepreneurs to implement mitigation strategies within their facilities: in recent years, many authors have demonstrated the financial return of improving factories environmental sustainability, something which has become a key factor for a company's long-term success (Cavallo et al., 2012; Despeisse et al., 2012; Longo & Mura, 2017; Stock & Seliger, 2016; Yang et al., 2017). The OECD (2011) advises that it should not only be big companies that undertake this path, as has been the case in the past, but also SMEs, which typically have more flexible business models and requirements, and therefore are easier and quicker in their ability to change.

**Resource and energy savings**. The most direct gain is linked with reduced operational costs due to more efficient processes and machinery, less energy and water demanding buildings, and optimised logistics. For example, when the building envelope is thermally insulated and old building technical systems are replaced with high-performing ones, the facility's overall energy demand decreases, and therefore operating costs are reduced. Many authors suggest that the rise in the cost of energy due to the economic crisis has significantly contributed to making companies more interested in energy efficiency, along with resource efficiency and pollution prevention (Dombrowski & Riechel, 2013; May et al., 2016).

In addition to reduced operational costs within already existing buildings, another direct reduction strategy might come from incorporating green building concepts early in the design phase, such as system thinking[1] and integrative processes[2] (U.S. GBC, 2014). These encourage the building system to be considered and analysed as a series of interconnected elements that can be mutually helpful, so that powerful synergies and effective integration among systems may be discovered from the pre-design stage and throughout the creative process. A lot of immediate benefits are available, such as reducing operational costs, discounting the initial costs of the construction phase (Ball et al., 2013; Pojasek, 2008). This entails optimising material utilisation during the construction stage, finding uncaptured energy reuse potential, or sizing systems fit-for-purpose. For example, if the integration of waste heat from industrial processes and the building's heating system is incorporated early in the design, a smaller and more efficient Heating, Ventilation and Air Conditioning (HVAC) system can be installed. Conversely, if the same synergy is identified during or after the building's construction, adjustments can still be put into place, resulting in lower running costs, but the price of the equipment does not reduce.

Moreover, information and smart technologies can help to find failures in the systems or uncaptured value, thus leading to reductions in costs, by means of modelling, automations, and commissioning. According to Yang et al. (2017), examples of uncaptured value exist in almost all companies, and the ineffective management of industrial waste streams is a recurrent one. Hence, since digital means can strongly support firms towards becoming energy efficient, the EU is pushing SMEs for a Twin Transition (Green and Digital) (The Twin Transition: How Can Green Growth and Digital Transformation Go Hand in Hand to Drive Europe's Recovery?, n.d.).

Although the long payback period of investments is a common obstacle in this field, it seems that a growing number of companies now accept longer return terms for greater environmental benefits (Addy, 2008). However, while "fast" returning measures—like envelope thermal insulation—are more easily applied on a voluntary basis, large-scale interventions may need adequate support from the top (i.e. institutions).

**Policy framework, regulations, and incentives**. Aside from direct reductions in costs, incentives and tax rebates resulting from complying or even exceeding regulations might encourage entrepreneurs to get sustainable factories. These kinds of rewards are essential in the transition to sustainable development. The 2030 Agenda for Sustainable Development represents a guiding force by which this shift can happen (United Nations, 2015). Several SDGs address manufacturing directly (9, 11,

---

[1] System thinking consists in considering the world, including the built environment, as a series of relationships in which all parts influence each other.

[2] Integrative process aims at supporting high-performance, cost-effective project outcomes through an early analysis of the interrelationships among systems.

12), implying factories should contribute to a clean, healthy and smart living environment. As a result, many strategies, plans, and measures have been implemented worldwide to support the achievement of these SDGs.

In Europe, considerable emphasis has been placed on energy and resource efficiency as strategies for tackling climate change, including decarbonisation, the use of renewable energy sources, and infrastructure modernisation (COM(2010) 2020—Europe 2020. A Strategy for Smart, Sustainable and Inclusive Growth, 2010). A new EU strategy for industry has been depicted, demanding greater commitment from members states, institutions, and especially industry itself (COM(2017) 479—Investing in a Smart, Innovative and Sustainable Industry. A Renewed EU Industrial Policy Strategy, 2017). More recently, the EU Green Deal called for a full and urgent mobilisation of industry because, in order to to achieve a climate neutral and circular economy by 2050—one of the eight goals—more than a generation is would have to pass (The European Green Deal. COM(2019) 640 Final, 2019). To this end, in March 2020 the Commission adopted the new EU Industrial Strategy that particularly addresses the twin challenge of the green and digital transformation of manufacturing (A New Industrial Strategy for Europe. COM(2020) 102 Final, 2020), along with a communication specifically targeted on SMEs (A SME Strategy for a Sustainable and Digital Europe. COM(2020) 103 Final, 2020). These also provide a basis to promote of a 'Renovation Wave' for a more sustainable built environment. As a result, specific measures have been implemented for the setting of requirements and thresholds that mainly focus on energy and resource efficiency in industry.

In this context, compliance with regulations is clearly the first and most important sustainability driver for entrepreneurs to prevent fines, penalties, and long delays in meeting requirements, not to mention customers choosing other brands when the environmental responsibility of companies is low (OECD, 2011).

On the other hand, policies wish to go beyond mere compliance and, at national and local levels, support is often given to exemplary companies that expand the boundaries of sustainability. This can be combined with a "carrot and stick approach" (incentives versus enforcement) to penalise inefficient organisations while rewarding those that are efficient; but no extensive applications have been made as of yet (May et al., 2016). Rather, most common supports are incentives, subsidies, and tax rebates. In addition to institutions, certain banks offer helpful conditions to support environmentally-friendly and efficiency interventions (Delpiano & Trombetti, 2011). Some local regulations also allow extra volume when eco-design requirements are met, as well as other structural and non-financial incentives including the chance to build extra floor space or the offer of expedited permissions (U.S. GBC, 2014). Others offer consultations with energy experts when implementing investments. In Europe, rewards particularly address reducing energy demand (e.g., retrofitting actions on the building envelopes and systems), increasing the use of renewable and clean energy (photovoltaic, solar heating, micro-wind and geothermal, biomass, Green and White certificates).

Nevertheless, such incentives remain focused mainly on the environmental dimension as the most relevant source of impact, but the opportunity to exploit reciprocal opportunities and social and perceptual aspects should also be addressed.

**Correlation between workers' well-being and productivity**. In this regard there are also interesting benefits from the increased productivity and fidelity of employees whose working environment is comfortable and pleasant. According to a World Green Building Council (World GBC) survey on pre- and post- occupancy in green offices, environmental health and well-being benefits relating to lighting, indoor air quality, interior layout, biophilia and views, acoustics, location and amenities can result in reduced absenteeism, medical complaints, staff turnover/retention, and increased revenue (Laski, 2018). This survey therefore suggests that an employee's satisfaction with the indoor comfort and outdoor amenities of a factory is important for the competitiveness of the firm itself.

**Green marketing and green reputation**. Going green not only saves money for the company in the short term, but it can also pay money in the medium-long term through green marketing. Indeed, the search for sustainability as a way to communicate excellence to the market frequently leads to the discovery of untapped value and missed opportunities. This is the basis of sustainable manufacturing and sustainable entrepreneurship: discovering and exploiting hidden value or opportunities for the company while also benefiting for the environment and the wider community (Patzelt & Shepherd, 2011; Pojasek, 2008). In practical terms, it entails repurposing traditional market failures (e.g., negative environmental externalities) into opportunities for the business by increasing its competitiveness. In a nutshell, this means taking advantage of certain company values to strengthen its value proposition in a sustainable manner.

Nowadays sustainability is "sexy" and in vogue for many big firms in relation to factory design (Addy, 2008). However, it is important to clarify that actual benefits are provided when green marketing as communication practice is related to a genuine effort within the activities of eco-design in the manufacturing process, facilities, product chains, and Corporate Social Responsibility. In this case, it can be a powerful means for repositioning a firm, differentiating it from comparable companies, and increasing its attractiveness (beyond offering a reduction in energy and costs). Otherwise, it is just green washing—that is, claiming to be environmentally friendly without being so—something which has been demonstrated to be an ineffective means of marketing beyond simply being misleading. In this regard, green building certifications like LEED, BREEAM, Green Star can help companies prove their sustainable performances are legitimate, as well as attracting longer lease terms and reducing the number of vacant factories (Green Building Council Australia, 2020).

**Territorial attractiveness**. Green marketing and enhanced competitiveness, in turn, can contribute to local development. In fact, in a virtuous circle, the territory in which an organisation operates can profit from having sustainable factories within its borders, by becoming more appealing to both visitors (e.g. wine and food tourism), capital, and new businesses (Cavallo et al., 2012; Papa et al., 2017).

In this context, the scenic or aesthetic value might be a valid asset or even a marketable landscape "product" (Bell, 2004). The same importance can be attributed to environmental value. Interestingly, according to Warren et al. (2005), when a construction excels in its environmental performance, locals are more willing to accept the facility in their landscape. They also illustrate that when good externalities are involved, such as surplus energy production for the neighbourhood, 'inverse NIMBY' syndrome can even occur: rather than saying Not In My Back Yard, residents may welcome the structure in the area as something to be proud of.

Some areas set local incentives for this very purpose, and thematic-driven developments are conceived to boost territorial marketing. For example, in the Emilia-Romagna region (Italy) there are several clusters that have created their own branding interlaced with sustainability and aesthetics, among others: the wellness valley, as well as a number of biotechnology, health, automotive, furniture and food sectors.

## Other Drivers

In addition to the broad range of economic drivers, there is a second category that deals with the increased awareness of more conscious consumers. There is evidence that stakeholders' and consumers' more acute sensitivity to ethical and environmental issues is progressively influencing the attitude of businesses (Despeisse, 2013; Qualizza, 2012; Yang et al., 2017).

The adoption of sustainability and green marketing practices is in fact supported by both consumer choices and policymakers, and this is reflected in the regulations. At EU policy level, the intent is to boost competitiveness in addition to sustainable growth (COM(2017) 479—Investing in a Smart, Innovative and Sustainable Industry. A Renewed EU Industrial Policy Strategy, 2017). At the same time, the Commission mentions the importance of awareness campaigns on end users to promote conscious and sustainable consumption (COM(2016) 739—Next Steps for a Sustainable European Future European Action for Sustainability 2016), which in turn can affect the weight that companies themselves assign to sustainability in production (Cavallo et al., 2012). The more the public becomes informed, the more companies are encouraged to be sustainable. In general, their focus is only on aspects that concern the environment.

Instead, since truly sustainable industrial buildings must meet best practice benchmarks across a range of categories (Green Building Council Australia, 2020), any aspect that improves a company's image can help to push the change forward. This is to argue that improving communication on business performance should not only involve the environmental image of the company, but it ought also to address its perceptual-aesthetic and social image, both of which impact the choice of consumers.

**Visual and perceptive image**. Although "beauty" has long been overlooked in factory design as an unnecessary consideration, it is now increasingly being proven to be important for a company's public image and competitiveness, as well as for territorial value (Busquets i Fabregas, 2007). The opposite is also true: a factory that is poorly built, uncomfortable, and not integrated into its surroundings can have a negative impact on a corporation's reputation. We are therefore gradually assisting in the re-appraisal of factory beauty and the social significance of the workplace at community level.

In fact, the value of a well-designed built environment in generating social and economic benefits is known (Macmillan, 2006). On the one hand, a pleasant factory can promote the corporation's image and even contribute to the area's identity, sense of place, and community attachment. On the other, a good design may allow the reduction of operational costs and increase the efficiency of both systems and employees. Nevertheless, once again, this topic requires a holistic approach. In the way that an efficient factory is not a synonym for a comfortable workplace, it is not sufficient to focus only the design of pleasant and multi-functional workplaces if one wishes to make a factory successful: as happens with every type of building, it is necessary to balance perceptual-aesthetic aspects with functional requirements and costs (Bell, 2004).

Let's take a very simple example focused on colours. Imagine the overall impact on the landscape of a factory whose roof is finished with light-coloured materials reflecting solar radiation (thus reducing the heat island effect and lowering the demand for cooling in the building); whose interiors are finished with bright, reflective paints that multiply the lighting level without dazzling employees (thus improving well-being, comfort, and productivity); and whose large façades are cladded with panels that make use of local palettes combined with graphic design to reduce the perception of massive volume, while expressing the company's values. Imagine that the result, combined with other strategies, is not only environmentally sound, but also so pleasant that local communities are proud of their living environment, factory included. Imagine if the driver of every car that passed by the factory became so impressed that afterwards, at the supermarket, when they see the wheat, for instance, that is produced by that firm, they decide to buy it. Then imagine a similar company that sees these benefits and decides to retrofit its own buildings as well. As a ripple effect, other target groups which could be positively impacted can be easily detected. In this way, the attractive and sustainable factory, which has a low environmental impact, a pleasing appearance, and is efficient and comfortable for workers, appears to be beneficial too (Calabrò, 2016).

It is critical to highlight that these actions on the architecture of factories must be consistent with the company's core values (Qualizza, 2012) in order to effectively promote the firm while having a positive impact on employees' well-being and productivity. In technical terms, they must be coherent with principles of "marketing aesthetic" (Schmitt & Simonson, 1997) to convey the firm's image and philosophy quickly and effectively, but also be truly sustainable.

The potential barrier to this approach is that it can be difficult to measure/quantify the added value of good design, and the risk of mere beautification or greenwashing practices is very high (Addy, 2008; Tandy, 1975).

**Social image**. Like for visual and perceptive qualities, there is an increasing correlation between social sustainability and financial gains. Companies which implement actions and policies to provide high-quality, differentiated services to employees, as well as public facilities to enrich the neighbourhood or even the larger area, are often rewarded with a stronger corporate image and competitiveness.

Customers are willing to pay for sustainable brands, and millennials especially want sustainable practices to be authentic rather than mere green or social-washing

(Rudominer, 2017). Mission communication, sustainability leadership and loyalty of employees seem to have a high impact on corporate financial success; a lower but still positive impact is due to improved social care and work life (Schönborn et al., 2019). The same authors prove that benefits are stronger when more than a single dimension is involved: once again, synergies among dimensions appears to be key.

**Responsibility**. No less important is the growing sense of responsibility for the many challenges the world is currently facing. Raising general awareness of the detrimental impact and of unsustainable factories might force the businessmen to take their role more seriously. This is beyond profit: it deals with the idea of being part of a collective movement towards a more sustainable, healthy, smart, and just living environment. As for behavioural changes in individuals there are many people who are aware of global challenges and act accordingly, the same is true among businessmen. What is certainly true is that the more a person understands their own impact, the more likely they are to take corrective action. Hence responsibility and awareness levels of awareness go hand in hand.

To summarise, a sustainable factory can provide both direct and indirect benefits to a company, such as lower operating and capital costs, access to public and private incentives, increased competitiveness through green marketing, social responsibility, and improved corporate image. As a result, a factory that is integrated into the landscape is also deemed economically beneficial.

Going one step further, Hermann et al. theorised the "positive impact factory" (Herrmann et al., 2015)—a factory that not only reduces environmental and social impacts, but even contributes to regenerating the surrounding area. The concept is very close to that of "regenerative design" promoted by the U.S. GBC (2014): that is, a project that is beneficial to the territory, for example, by providing energy to close neighbours during the night; offering local sewage treatment facilities; creating new habitats by restoring lakes and/or building greenhouses for living organisms and new species; and offering services to the local community such as sport and childcare facilities.

This clearly shows why territories themselves have a vested interest in hosting and supporting sustainable and socially minded factories. It has been demonstrated that the territory's value increases and that new capital and innovative firms are attracted to the area. Furthermore, sustainable factories might set off a virtuous cycle of restoration involving local facilities and competitors. As a result, sustainable development at the local level can be fostered while also contributing to the achievement of global goals.

In this framework consumers' perception of companies has emerged as a critical driver in changing companies' behaviour. The more aware they become of global challenges in terms of environmental and social issues, the more demands they make of firms to act consistently in terms of environmental performance and physical features of the facility. Holistic and genuine green responsibility is thus required of firms.

Obviously, none of the described drivers would be a cure-all by themselves, but a combination of them could help to accelerate change.

## Means to Support the Design of Sustainable Factories

Now that potential drivers for designing or renovating industrial facilities so that they would be in harmony with the landscape have been presented, it is time to get to know how to act. Given the lack of accurate awareness of negative impact among businesses, as well as the scarcity of information on feasible measures to reduce the effects and their associated advantages, providing businesses with tools capable of guiding them through the design process could be a good starting point.

Based on what has been discussed up to this point, businesses need to be informed about their negative impact on the landscape and about possible corrective actions they could implement. Without knowing what the impact is and what can be eliminated, compensated for, or mitigated, businesses will act ineffectively. Moreover, they need to be aware of the potential benefits that a sustainable factory might generate. Not to mention the capability of firms to communicate their performance to the market.

For these reasons, sustainability assessment could help since it has been recognised as crucial to the dissemination of good practices in regards to the built environment (Ding, 2008). Hence, tools measuring both the negative and positive effects of industrial sites on the landscape have been developed. At the same time, guidelines for the mitigation of the negative effects have been disseminated.

However, most of these tools are sectorial, focusing only on a single part of the problem at a time. In particular, when it comes to assessing the impact of sustainable manufacturing at plant or facility level, the available tools can be divided into two categories: environmental and aesthetic-perceptual. Social effects are frequently included within both categories as an indirect consequence.

**Environmental assessment tools**. Methods and instruments to measure the environmental impact of manufacturing have a large number of contributions and applications, in both the scientific literature and the regulatory environment. Academics and industry have developed several tools and strategies to improve the sustainability of manufacturing systems whilst extracting tangible benefits for firms (Ball et al., 2013). In particular, decision-support procedures capable of jointly considering the plurality of elements that interact with each other are increasingly being used in environmental impact assessments, often based on multi-criteria analyses. Frequently referred to as Rating Systems (RSs), they are tools that simplify the assessment of a complex issue without losing its interrelations. Not only are RSs useful in finding the critical features of a problem, but they are also a good way of communicating environmental performances (Jesinghaus, 2000).

Well-established tools in this field are Green Building Rating Systems (GBRSs). These are third-party, voluntary, and market-driven standards that measure buildings' sustainability by multi-criteria assessments, whilst and encouraging the adoption of environmentally, socially and economically sustainable practices in the design, construction and operation of buildings (Marchi et al., 2021). For a decade many Green Building Councils have developed schemes suitable for factories, usually adapting procedures designed for other structures such as offices or warehouses. LEED, BREEAM, CASBEE, are among the most common worldwide (Kaur et al., 2016). They assess the sustainability level of a construction, at site and building levels, by adding up indicators from various topics like location and transportation, water management, site quality, energy efficiency, indoor environmental quality, and material utilisation. They are usually adopted during the early design stage (whether new construction or renovation), when the company is guided by the design team to select strategies and practices that provide the greatest environmental benefits. Thus, beyond serving as an assessment procedure and design support tool, GBRSs also have an important part to play in communicating to companies the benefits of their adoption. However, these tools are often criticised for being time-consuming (for the data retrieval and calculations of single indicators), as well as for the complexity of some indicators or their inaptness in regard to specific industrial sectors, which prevents the widespread adoption of GBRSs in the field. Chen et al. (2013) advocate for their simplification in order to achieve maximum dissemination and implementation.

Similarly, few instruments have been devised that combine site, building and process levels. It is the case of the Organisation for Economic Co-operation and Development (OECD) Sustainable Manufacturing Toolkit (OECD, 2011) that aims to provide companies around the globe with a series of practical steps to improve the efficiency of their production processes and products at the plant level. Some other tools have been developed by academics (Davé et al., 2016; Despeisse et al., 2016; Fantini et al., 2015; Provincia di Bologna, 2008) who propose their own assessment frameworks and methodologies in attempt to be comprehensive.

Despite this effort, a crucial gap is still detected in all these instruments: there are no tools on the market that cover the analysis of all levels of a manufacturing site, identify critical points, and suggest corrective actions at the same time (Despeisse, 2013). The efficiency of processes is usually neglected from the assessment of the performance of industrial buildings, even though they frequently share the same energy infrastructure. Thus the chance to positively exploit the interrelationships is lost (Ball et al., 2013). As a result, important opportunities are missed, and there is also the risk of grading something as sustainable that is not, because the triple bottom line of sustainability is not considered as a whole. This means, theoretically, that a company with a LEED Platinum facility—the highest grade for that GBRS—can be considered "sustainable" even if it exploits workers or contrasts heavily with its surroundings. A similar thing could also happen to a corporation with an excellent Global Reporting Initiative index (GRI) that focuses on the process but has a poorly planned and bad performing facility. For this reason, Akter (2016) advises that one should look carefully for the evaluation criteria when the term "green factory" is used.

Moreover, in contrast to what has been discussed so far in the book, most environmental assessment systems for factories do not evaluate the other landscape characteristics.

**Aesthetic and perceptive assessment tools.** On the other hand, although industrial sites are generally among the types of constructions that have the greatest sway on the visual appreciation of the landscape (Farjon et al., 2009), little attention has been paid to them in impact assessment procedures. Until now, few guidelines and procedures have specifically addressed aesthetic and perceptive interferences of industry, and these generally focus only on the visual aspect. The methodology adopted by these procedures is similar to that of environmental impact assessment tools: a set of indicators is used to gauge the impact (mostly qualitative rather than quantitative), which is then followed by a list of potential mitigation measures.

Most of these tools have been developed at the policy and planning level. Some will be considered as a means of example. A study commissioned by Milano province (Italy) in 2005 that was attached to the updated Territorial Plan is particularly noteworthy (Provincia di Milano, 2012) because the impact of industrial sites on the landscape was addressed using evaluation metrics that linked different types of transformation to distinct landscape types, progressing through multiple scales and eventually arriving at a series of possible mitigation interventions. Another study, conducted by the Polytechnic of Torino for the Sardegna region (O.P.U.Q.P., 2015), demonstrates how aesthetic-perceptual features have started to gain attention in regard to industrial facilities. The study identifies recurrent detrimental effects of factories in the region and proposes mitigation strategies for each type of operation, including both environmental and perceptive aspects. On the same wavelength is the work of other Italian regions such as Emilia-Romagna and Puglia. Some other useful guidelines resulted from European-funded projects, such as the Interreg III Medocc PAYSMED (Busquets i Fabregas, 2007) that proposed a series of mitigation strategies for industrial polygons proceeding from general site to a more detailed level.

Overall, the number of guidelines and tools for the assessment of the impact on landscapes impact by factories is limited, so other interesting references can be retrieved from similar sectors and constructions. It is certainly the case for energy infrastructure, whose integration into the landscape has been subjected to plenty of studies and projects (Busquets i Fabregas, 2007; Devine-Wright & Howes, 2010; Ginelli & Daglio, 2016; Maehr et al., 2015; Mueller et al., 2017; O.P.U.Q.P., 2015; Wolsink, 2007). The topic of farm buildings and their integration into the rural landscape has been equally addressed (García et al., 2003; O.P.U.Q.P., 2015; Rosley et al., 2013; Tassinari et al., 2013).

However, despite being stimulating, these instruments are often limited to specific geographic locations, while a more widely applicable instrument (like GBRSs but for the perceptual dimension) is still unavailable. Furthermore, unlike GBRSs, these procedures frequently entail the subjective opinion of experts, for which they are criticised for not being objective, and the results are subsequently difficult to compare to each other. Moreover, as mainly institutional-promoted guidelines, they are not market-driven, so they lack the communication of performance component that might encourage uptake by entrepreneurs.

To sum up, it appears that these tools are useful but limited. There is no instrument currently available that can address the problem of a 360-degree sustainable factory as intended in this book, hence the adoption of a multi-scale and multi-faceted approach to the issue. Instead, a broad perspective capable of recognising connections and mutual links across the numerous aspects of the system must be used to identify positive synergies and eradicate perception of sustainability as an extra cost. Furthermore, in order to maximise implementation, these tools should be easy to understand and possibly be able to be applied by businesses themselves, as well as capable of returning insightful results that can be used to communicate with the market.

## References

A New Industrial Strategy for Europe. COM(2020) 102 final, (2020). https://ec.europa.eu/info/sites/info/files/communication-eu-industrial-strategy-march-2020_en.pdf.

A SME Strategy for a sustainable and digital Europe. COM(2020) 103 final, (2020). https://eur-lex.europa.eu/legal-content/EN/TXT/PDF/?uri=CELEX:52020DC0103&from=EN.

Addy, R. (2008). Green is the goal. In *FOOD manufacture, Ottobre* (pp. 45–46). https://www.foodmanufacture.co.uk/Article/2008/09/29/Green-is-the-goal.

Akter, A. (2016). How green are the 'green factories'? In *Textile today*. https://www.textiletoday.com.bd/green-green-factories/.

Ball, P. D., Despeisse, M., Evans, S., Greenough, R. M., Hope, S. B., & Kerrigan, R., et al. (2013). Factory modelling: Combining energy modelling for buildings and production systems. *IFIP Advances in Information and Communication Technology, 397*(PART 1), 158–165. https://doi.org/10.1007/978-3-642-40352-1_21.

Bell, S. (2004). *Elements of visual design in the landscape* (2nd ed.). Spon Press.

Busquets i Fabregas, J. (Ed.), (2007). *Per una corretta visione del paesaggio: linee guida* (Traduzione). Generalitat de Catalunya. http://www.paysmed.net/.

COM(2010) 2020—Europe 2020. A strategy for smart, sustainable and inclusive growth, (2010). https://ec.europa.eu/.

COM(2016) 739—Next steps for a sustainable European future European action for sustainability, (2016). https://ec.europa.eu/transparency/regdoc/?fuseaction=list&coteId=1&year=2016&number=739&version=ALL&language=it.

COM(2017) 479—Investing in a smart, innovative and sustainable Industry. A renewed EU Industrial Policy Strategy, (2017). https://eur-lex.europa.eu/legal-content/EN/TXT/?uri=COM:2017:0479:FIN.

Calabrò, A. (2016). La "fabbrica bella": architettura e sostenibilità. Il vantaggio d'essere più produttiva e competitiva. *HuffingtonPost Italia*. http://www.huffingtonpost.it/antonio-calabro/sviluppo-sostenibile-made-in-italy-aziende-italiane-_b_10040560.html.

Cavallo, M., Degli Esposti, P., & Konstantinou, K. (Eds.), (2012). *Green Marketing per le aree industriali. Metodologie, strumenti e pratiche*. Franco Angeli.

Chen, D., Schudeleit, T., Posselt, G., & Thiede, S. (2013). A state-of-the-art review and evaluation of tools for factory sustainability assessment. *Procedia CIRP, 9*, 85–90. https://doi.org/10.1016/j.procir.2013.06.173.

Davé, A., Salonitis, K., Ball, P., Adams, M., & Morgan, D. (2016). Factory eco-efficiency modelling: Framework application and analysis. *Procedia CIRP, 40*, 214–219. https://doi.org/10.1016/j.procir.2016.01.105.

Delpiano, A., & Trombetti, G. (Ed.), (2011). *EnergiApea. Vincere la sfida energetica* (RicercaeProgetto—Galassi Mingozzi e associati (Ed.)). Tipografia metropolitana Bologna.

Despeisse, M., Ball, P. D., Evans, S., & Levers, A. (2012). Industrial ecology at factory level—a conceptual model. *Journal of Cleaner Production, 31*, 30–39. https://doi.org/10.1016/j.jclepro.2012.02.027.

Despeisse, M., Davé, A., Litos, L., Roberts, S., Ball, P., & Evans, S. (2016). A collection of tools for factory eco-efficiency. *Procedia CIRP, 40*, 542–546. https://doi.org/10.1016/j.procir.2016.01.130.

Despeisse, M. (2013). *Sustainable manufacturing tactics and improvement methodology. A structured and systematic approach to identify improvement opportunities*. School of Applied Science, Cranfield University. https://doi.org/10.1017/CBO9781107415324.004.

Devine-Wright, P., & Howes, Y. (2010). Disruption to place attachment and the protection of restorative environments: A wind energy case study. *Journal of Environmental Psychology, 30*(3), 271–280. https://doi.org/10.1016/j.jenvp.2010.01.008.

Ding, G. K. C. (2008). Sustainable construction-the role of environmental assessment tools. *Journal of Environmental Management, 86*, 451–464. https://doi.org/10.1016/j.jenvman.2006.12.025.

Dombrowski, U., & Riechel, C. (2013). Sustainable factory profile: a concept to support the design of future sustainable industries. In *11th Global Conference on Sustainable Manufacturing* (pp. 72–77).

Fantini, P., Palasciano, C., & Taisch, M. (2015). Back to intuition: Proposal for a performance indicators framework to facilitate eco-factories management and benchmarking. *Procedia CIRP, 26*, 1–6. https://doi.org/10.1016/j.procir.2014.07.099.

Farjon, H., Van Der Wulp, N., & Crommentuijn, L. (2009). Programa de seguiment de la percepció i l'apreciació del paisatge als Països Baixos. In *Plecs de Paisatge*. Observatori del Paisatge.

García, L., Hernández, J., & Ayuga, F. (2003). Analysis of the exterior colour of agroindustrial buildings: A computer aided approach to landscape integration. *Journal of Environmental Management, 69*(1), 93–104. https://doi.org/10.1016/S0301-4797(03)00121-X.

Ginelli, E., & Daglio, L. (2016). Le infrastrutture per le energie rinnovabili nel paesaggio. Strumenti di progetto e traiettorie dell'innovazione. *TECHNE, 11*, 119–126. https://doi.org/10.13128/Techne-18410.

Green Building Council Australia. (2020). *Green star in focus. The case for sustainable industrial buildings*. https://gbca-web.s3.amazonaws.com/media/documents/green-star-in-focus-the-case-for-sustainable-industrial-buildings.pdf.

Herrmann, C., Blume, S., Kurle, D., Schmidt, C., & Thiede, S. (2015). The positive impact factory—transition from eco-efficiency to eco-effectiveness strategies in manufacturing. *Procedia CIRP, 29*, 19–27. https://doi.org/10.1016/j.procir.2015.02.066.

Jesinghaus, J. (2000). On the art of aggregating apples and oranges. In *EVE workshop*.

Kaur, G. P., Gupta, P., & Syal, M. (2016). Adoption of green practices in industrial buildings: An action research on capacity building of stakeholders towards green factories. *International Journal of Sustainable Land Use and Urban Planning, 3*(2), 1–12.

Laski, J. (2018). Doing right by planet and people. In *World green building Council*. https://www.worldgbc.org/better-places-people%0A.

Longo, M., & Mura, M. (2017). Assessing sustainability within organizations: The sustainability measurement and management lab (SuMM). In G. Campana (Ed.), *Sustainable design and manufacturing 2017* (vol. 68, pp. 339–346). Berlin: Springer. https://doi.org/10.1007/978-3-319-57078-5.

Macmillan, S. (2006). Added value of good design. *Building Research & Information, 34*(3), 257–271. https://doi.org/10.1080/09613210600590074.

Maehr, A. M., Watts, G. R., Hanratty, J., & Talmi, D. (2015). Emotional response to images of wind turbines: A psychophysiological study of their visual impact on the landscape. *Landscape and Urban Planning, 142*, 71–79. https://doi.org/10.1016/j.landurbplan.2015.05.011.

Marchi, L., Antonini, E., & Politi, S. (2021). Green building rating systems (GBRSs). *Encyclopedia, 1*(4), 998–1009. https://doi.org/10.3390/encyclopedia1040076.

May, G., Stahl, B., & Taisch, M. (2016). Energy management in manufacturing: Toward eco-factories of the future—a focus group study. *Applied Energy, 164*, 628–638. https://doi.org/10.1016/j.apenergy.2015.11.044.

Meadows, D. (1999). *Leverage points: places to intervene in a system*. https://doi.org/10.1080/02604020600912897.

Mueller, C. E., Keil, S. I., & Bauer, C. (2017). Effects of spatial proximity to proposed high-voltage transmission lines: Evidence from a natural experiment in Lower Saxony. *Energy Policy, 111*(2017), 137–147. https://doi.org/10.1016/j.enpol.2017.09.023.
O.P.U.Q.P. (2015). *Linee guida per i paesaggi industriali in Sardegna.*
OECD. (2011). *OECD sustainable manufacturing tookit—Start-up guide.* www.oecd.org/innovation/green/toolkit.
Pal, U. (2002). Identifying the path to successful green manufacturing. *JOM Journal of the Minerals Metals and Materials Society, 54*(5), 25. https://doi.org/10.1007/BF02701691.
Papa, R., Gargiulo, C., Russo, L., & Franco, S. (2017). On the relationship between the promotion of environmental sustainability and the increase of territorial competitiveness: The Italian case. *International Journal of Sustainable Development and Planning, 12*(4), 655–666. https://doi.org/10.2495/SDP-V12-N4-655-666.
Patzelt, H., & Shepherd, D. A. (2011). Recognizing opportunities for sustainable development. *Entrepreuneurship Theory and Practice, 35*(4), 631–652.
Pojasek, R. B. (2008). When is green manufacturing green? *Manufacturing Engineering, 141*(2), 18–19. www.sme.org/manufacturingengineering.
Provincia di Bologna. (2008). *APEA - Aree Produttive Ecologicamente Attrezzate. Linee guida* (M. Cavallo (Ed.)).
Provincia di Milano. (2012). *Adeguamento del PTCP alla LR 12/2005. Repertorio delle misure di mitigazione e compensazione paesistico-ambientali* (pp. 1–193). Piano Territoriale di Coordinamento Provinciale.
Qualizza, G. (2012). Inventare lo spazio di una green factory. In M. Cavallo, P. Degli Esposti, & K. Konstantinou (Eds.), *Green marketing per le aree industriali. Metodologie, strumenti e pratiche* (pp. 220–238). Franco Angeli.
Rosley, M. S. F., Lamit, H., & Rahman, S. R. A. (2013). Perceiving the aesthetic value of the rural landscape through valid indicators. *Procedia—Social and Behavioral Sciences, 85*, 318–331. https://doi.org/10.1016/j.sbspro.2013.08.362.
Rudominer, R. (2017). *Corporate social responsibility matters: Ignore millennials at your peril.* https://www.huffpost.com/entry/corporate-social-responsi_9_b_9155670?guccounter=1&guce_referrer=aHR0cHM6Ly9teWZvb2Rzb2Z0d2FyZS5jb20v&guce_referrer_sig=AQAAAJxt7Ikv_vhrW86h7Lws_4S7clA537YqaxHfJ9vUQurYtWaknqsEuKNF9sll5DgPVtOc7Tu_uAgpDrTJi3ny5NgtFTM_IN3w2as6.
Schmitt, B., & Simonson, A. (1997). *Marketing aesthetics. The strategic management of brands, identity and image.* The Free Press.
Schönborn, G., Berlin, C., Pinzone, M., Hanisch, C., Georgoulias, K., & Lanz, M. (2019). Why social sustainability counts: The impact of corporate social sustainability culture on financial success. *Sustainable Production and Consumption, 17*, 1–10. https://doi.org/10.1016/j.spc.2018.08.008.
Stock, T., & Seliger, G. (2016). Opportunities of sustainable manufacturing in industry 4.0. *Procedia CIRP, 40*, 536–541. https://doi.org/10.1016/j.procir.2016.01.129.
Tandy, C. (1975). *Landscape of industry.* Leonard Hill Books.
Tassinari, P., Torreggiani, D., Benni, S., & Dall'Ara, E. (2013). Landscape quality in farmyard design: An approach for Italian wine farms. *Landscape Research, 38*(6), 729–749. https://doi.org/10.1080/01426397.2012.746653.
The twin transition: How can green growth and digital transformation go hand in hand to drive Europe's recovery? (n.d.). Retrieved August 3, 2021, from https://events.euractiv.com/event/info/the-twin-transition-how-can-green-growth-and-digital-transformation-go-hand-in-hand-to-drive-europes-recovery.
The European Green Deal. COM(2019) 640 final, (2019). https://eur-lex.europa.eu/legal-content/EN/TXT/?uri=CELEX:52019DC0640.
U.S. GBC. (2014). *LEED core concepts guide.* U.S. Green Building Council.

United Nations. (2015). *Transforming our world: The 2030 agenda for sustainable development (A/RES/70/1)*. https://documents-dds-ny.un.org/doc/UNDOC/GEN/N15/291/89/PDF/N1529189.pdf?OpenElement.

Warren, C. R., Lumsden, C., O'Dowd, S., & Birnie, R. V. (2005). "Green on green": Public perceptions of wind power in Scotland and Ireland. *Journal of Environmental Planning and Management, 48*(6), 853–875. https://doi.org/10.1080/09640560500294376.

Wolsink, M. (2007). Wind power implementation: The nature of public attitudes: Equity and fairness instead of "backyard motives." *Renewable and Sustainable Energy Reviews, 11*(6), 1188–1207. https://doi.org/10.1016/j.rser.2005.10.005.

Yang, M., Evans, S., Vladimirova, D., & Rana, P. (2017). Value uncaptured perspective for sustainable business model innovation. *Journal of Cleaner Production, 140*, 1794–1804. https://doi.org/10.1016/j.jclepro.2016.07.102.

# Chapter 4
# Pioneer Firms from the Agri-Food Industry

**Abstract** The case of agri-food is exemplary as it has turned sector-specific weaknesses into opportunities for many decades. The major impact on the environment due to its high-consumption processes, as well as frequent and intense contrasts with the rural landscape, clashes with the sector's core values of respecting nature and its inhabitants. This is strongly related with the architecture of the factory and surrounding facilities, and the food and beverage industry has understood, prior to many other sectors, the valuable opportunities that an exemplary factory can create. Not only has the relevance of the subject been recognised in the scientific literature, but also in practice several companies have accepted the challenge. This chapter presents six trends that recur in good agri-food firms around the world, retrieved from a structured examination of case studies. One interesting approach is, for example, to design underground buildings to both lower the visibility of the structure and the demand for space cooling by taking advantage of the ground's thermal properties. Although they are focused on the food and beverage industry, the presented patterns are easily adaptable to other industrial sectors.

**Keywords** Colour design · Food and beverage · Green factory · Underground storage · Wineries

## The Case of Agri-Food: From Weaknesses to Strengths

Sustainable factories entail a series of opportunities and benefits for the environment, society, and the businesses themselves. As discussed in Chap. 3, this has been proven by researchers, supported by policy makers, driven by consumers behaviour, and, not of minor importance, gradually understood by companies.

Among industrial sectors, agri-food has been a pioneer in terms of successful factory design for several decades since it started to turn certain environmental and social issues into business opportunities. Because of its highly energy-intensive and water-demanding operations, as well as its sharp contrasts with the rural settings in which its facilities are frequently located, the food and beverage sector is one of the industries with the most severe impact on the landscape. Furthermore, the sector's peculiarities require consistency between the factory's image and the corporation's



L. Marchi, *Designing Sustainable Factories*, Advances in Global Change Research 72,
https://doi.org/10.1007/978-3-030-92227-6_4

values, because the quality of goods for human consumption is inextricably linked to the quality of the area in which they are produced. As a result, landscape quality is essential for these firms, an increasing number of which are applying the same architectural quality generally reserved for headquarters to their facilities.

Hence, the case of food and beverage companies provides a tangible, real-life example of the idea that a sustainable factory is beneficial it is not something which has only been discussed in academia.

Before analysing the question in detail, a definition of the sector could be useful in order to frame the discussion properly. The food, or agri-food industry is defined as a group of businesses dedicated to the production of beverages or other finished and semi-finished products through the processing and transformation of raw materials derived from primary activities such as agriculture, animal husbandry, forestry, and fishing, with the goal of consumption by humans or animals (Industria Alimentare, 2012). The variety of processes involved makes generalisation difficult, but its major segments comprise (Clark, 2005): cereals, meat, fruit and vegetables, beverages, food ingredients, and dairy. Alternatively, from a technological point of view, these industries can be classified as: extractive (flour, sugar, oil); fermentative (wine, beer, cheese); canning and based on physical and mechanical operations (jam, sweets, coffee).

In many countries, this sector is still one of the most important in numerical and financial terms. In Europe, the food industry is made of SMEs (99%), the vast majority of which are in the South, where it represents one of the leading economic sector (Enea, 2016).

The impact of these companies on the environment, the scenery, and society is as important as their economic significance.

**Highly resources and energy consuming processes**. From a life-cycle perspective, the interactions of an agri-food company with the environment span the entire production chain, from raw material procurement to the storage and distribution of finished goods.

Agricultural resources are extremely vulnerable to climate change, atmospheric particulate deposits, and meteorological events such as droughts and floods. Thus availability, quality, and prices of agricultural ingredients are largely dependent on local weather conditions, which in turn affect the entire production and value chain of food industry (Fedriga, 2016). For this reason, the issue is genuinely considered by businesses in the sector.

In addition, agri-food operations frequently necessitate huge amounts of energy and resources, as well as involve transformation processes that threaten the balance of natural ecosystems. The largest consumer are involved in food conservation and disinfection, which require thermal treatments such as pasteurisation, sterilisation, drying, refrigeration and freezing (Lee & Okos, 2011). To generate thermal energy, companies usually employ steam boilers and compression refrigeration units that are powered by fossil fuels (Philipp et al., 2017). So, energy efficiency and the indoor thermo-hygrometric balance of the buildings, as well as renewable energy sources and efficient systems may significantly contribute to reducing the impact.

Beyond energy use, water demand is also significant: thousands of litres of fresh water per hour are utilised in these facilities. This resource is used both directly as an ingredient in the preparation of food and beverages, and indirectly as a means of production, such as in washing plants and steam generators (Lee & Okos, 2011). Even for this secondary purpose, current regulations do not often permit the water to be reutilised. Thus, water resource management—from supply to disposal—is economically critical for food firms, representing an important item of expenditure (Bortolini et al., 2017).

As a result, many firms recognised the necessity of implementing improvements and innovations targeting the reduction of water demands: this is to be compliant with upcoming requirements to limit primary water demands, as well as to adopt good practices for the treatment and recovery of water used in industrial processes.

**Contrast with the scenery**. Aside from environmental concerns, the aesthetic-perceptual and social interference of food and beverage factories with their surroundings is a crucial issue too. Beyond the poor architectural quality and lack of relationship with the scenery that these facilities share with those from other sectors, their peculiar location frequently accentuates the contrast. Many of these facilities are within or are close to rural areas, due to historic functional and logistical reasons, such as the perishable nature of food or the requirement for large areas of land for the storage of seasonal crops. In the countryside, where the density of construction is low and vegetation (cultivated or natural) dominates the landscape, the contrast between factories and their surrounding is more likely to be intense. Therefore large, grey, anonymous buildings, out of scale, standing alone within green surroundings, are often perceived as detracting from the landscape. It goes without saying that the more delicate the balance of the scenery, the more acute the interference.

This is certainly the case for many Mediterranean landscapes where, according to a survey conducted as part of the European project PAYS.MED.URBAN, industrial settlements are often responsible for the degradation of fringe agricultural landscapes (Regione Lombardia, 2011).

**Relationship between quality of products and place of production**. The two preceding points are tightly linked to the fact that there is a close and socially shared relationship between processing of edible goods and the quality of the environment from which they come from, which provides a significant portion of the product's value, both real and symbolic (e.g., healthiness, typicality, genuineness). Proof of this may be found in the claims of some internationally known Italian firms, such as "Good for nature" by Orogel,[1] or "The nature of first-hand" by Valfrutta.

The connection with nature is vital to these businesses, and in many cases the advertisement refers to an idyllic picture where men, crops and factories are in a balanced and peaceful relationship.[2] As a result, more than in any other sector, the

[1] The original claim in Italian is "Buono per natura", meaning that the product is tasty and healthy since it comes directly from nature. Its English translation even evokes another intriguing point of view: the product is good *for* the nature itself.

[2] The specific reference is to the long-lasting advertisement style of internationally-known Italian firms, for its industrial bakery products, produced within water mills (i.e., Barilla—Mulino Bianco).

harmonization of the industrial site with the landscape is crucial as further evidence of coherence with the firms' core values (e.g., sustainability, respect for nature, territorial valorisation). Otherwise a clear conflict emerges when the consumer, who is increasingly informed and interested in sustainability, finds themselves in front of a massive, automated warehouse where those much-lauded natural, healthy, and genuine products are produced or stored. This is also true of a building that obstructs the view of the countryside, is surrounded by red-hot asphalted areas, finished by grey aluminium insulated prefabs, and other features that in no way recall that naturalness, seasonality, and respect for the environment.

However, other sectors are quite different. Let's take as an example the automotive industry that is often associated with the values of innovation, technology, and speed: in this case, it has little to do with the "natural" landscape, and, the consistency of buildings with the context is not perceived as critical as it is for the food and beverage industry.

Thus, agri-food facilities have a very close while fragile relationship with the landscape.

## Recurrent Patterns in Exemplary Food and Beverage Factories

During the preceding decades, many companies have transformed the aforementioned critical attributes into growth opportunities. Since the functional requirements of proximity to the resource (e.g., crops, livestock) has weakened over this time, the suitability of the location of agro-industrial plants has necessitated a rethink. New relationships of exchange were developed with the landscape, turning the factory's impact into a positive, whilst simultaneously enabling the factory to take advantage of the quality of its surroundings. Thus, the sector's privileged but difficult relationship with nature became the trigger and driver to design more sustainable factories.

The success of the wine sector is a particularly meaningful example in this regard. It proves that the space and place of production increasingly matter for marketing (Giovanardi & Lucarelli, 2018) and especially when it comes to industrial tourism. Here, environmental, visual, and perceptive features of the landscape represent both inputs of the value chain and factors of identity for the product (Gandolfi, 2011). When the phenomenon is extended beyond the single product/firm to a territorial brand, the success is exponential. The well-known examples of the Chianti Valley in Tuscany (IT), or the wider Made in Italy brand give important insights into this subject.

Many impressively designed and sustainable wineries all around the world have been pioneers of this trend (Harea & Eplényi, 2017; Tassinari et al., 2013; Torquati et al., 2015). While, for the food industry, significant progresses has been made, but some lag is still detected when taking into consideration the numerous opportunities the sector could offer.

In general, the same barriers discussed in Chap. 2 continue to hamper many SMEs from changing their approach. However, these are more likely influenced by consumers and the local population when it comes to locally rooted companies. Hence, consumer loyalty, which depends highly on the company's social and environmental reliability and responsibility, represents a real driving force, and several companies are taking steps towards achieving corporate image coherence and enhancement while also being environmentally conscious. In this framework, certain territorial marketing projects and industrial tourism experiences have found significant development prospects in the food sector for the benefit of both firms and the area (Fanfani & Pieri, 2016; Tassinari et al., 2013).

Therefore, given that the agri-food sector seems to have had a piloting role in designing more sustainable factories with a holistic perspective, a list of exemplary factories has been compiled. This is part of a research project whose overall methodology and outcomes will be better outlined in Chap. 5. A number of notable food and beverage industrial sites have been selected, particularly those that have adopted good mitigation practices at both the physical level (e.g., open space, shape and volume, technology) and management level (e.g., services for workers). Eligibility criteria were, namely, that the factory should have been built or retrofitted after 2000, and that the best practices were related to at least one of the three landscape dimensions, i.e. with the aim of reducing the detrimental effect of the factory on the environment; the lessening of aesthetic-perceptual interferences with the surroundings; and/or promoting the well-being of workers and local communities. A more detailed explanation of the methodology for the selection and analysis of case studies is provided in Marchi (2020).

**Annex 1** reports brief info-sheets for each selected case, along with references for further details. From the forty-five selected factories, a series of recurrences have been detected, proving that virtuous approaches and practices to sustainable constructions are spreading within agri-food firms around the world. Figure 4.1 illustrates an extract of the procedure to analyse these cases, from which, in turn, six trends have been derived and will be discussed below.

From this list, a different approach emerges between the retrofitting of existing factories and the construction of new ones. For the former, it is frequent to resort to the surface treatment of the factory image with little or no association with good practices to reduce the impact on the ecosystem. For new and more recent factories this divergence instead seems to be reduced: the settlement and architectural quality of the factory, in an aesthetic sense, is often integrated with the construction quality and good environmental practices that involve the site, the building, and the process acting in synergy.

In general, it is observed that many good practices target energy efficiency and renewable energy production, partly due to the number of policies and incentives offered in the last few decades. Enhancement of corporate image and morphological harmonisation of facilities with the surrounding environment also appear to be mainstream (more than half of the sample exhibit these benefits), demonstrating once again the relevance of the issue for the sector.

The analysis has also pointed out that food companies are particularly interested in the environmental quality of the site and, when possible, they prefer to settle in existing areas rather than consuming new soil. Beverage industries also appear interested in site quality and energy performances, but they further focus on daylight

| Code | Case study | Country | Period | Integrated Process | Location and Transportation (LT) | LEED for Neighborhood Development Lo | Sensitive Land Protection | High Priority Site | Surrounding Density and Diverse Uses | Access to Quality Transit | Bicycle Facilities | Reduced Parking Footprint | Green Vehicles | Sustainable Site (SS) | Site Assessment | Site Development - Protect or Restore H | Open Space | Rainwater Management | Heat Island Reduction | Perceptual and Aesthetic Aspects | Morphological Harmonisation | Colour and Material Harmonisation | Noise Pollution | Site Accessibility and Receptiveness | Work Space and Leisure Area Design | Consideration of Symbolic Character | Corporate Image | Rate of active credits | Percentage |
|---|---|---|---|---|---|---|---|---|---|---|---|---|---|---|---|---|---|---|---|---|---|---|---|---|---|---|---|---|---|
| A | **Food industry** | | | | | | | | | | | | | | | | | | | | | | | | | | | | |
| 1 | Melinda | IT | 2014 | | 2/8 | | | ■ | | | | ■ | | 2/6 | | | ■ | | ■ | 3/7 | | ■ | | | ■ | | ■ | 17/49 | 35% |
| 2 | Bio Pastificio IRIS | IT | 2015 | | 0 | | | | | | | | | 3/6 | | | ■ | ■ | ■ | 3/7 | | | | ■ | ■ | | ■ | 15/49 | 31% |
| 3 | Almazara Olisur Oli | CL | 2009 | | 0 | | | | | | | | | 0 | | | | | | 3/7 | ■ | ■ | | | | ■ | | 8/49 | 16% |
| 4 | Stabilimento Maina | IT | 2015 | | 1/8 | | | ■ | | | | | | 0 | | | | | | 1/7 | | | | | | | ■ | 2/49 | 4% |
| 5 | Grandi Salumifici It | IT | 2013 | | 0 | | | | | | | | | 0 | | | | | | 1/7 | | | | | | | ■ | 3/49 | 6% |
| 6 | Solare Manufaktur | DE | 2012 | ■ | 1/8 | | | | | | | | ■ | 0 | | | | | | 1/7 | | | | | | | ■ | 8/49 | 16% |
| 7 | Kerrygold | UK | 2008 | | 1/8 | | | ■ | | | | | | 0 | | | | | | 0 | | | | | | | | 4/49 | 8% |
| 8 | Pastificio Felicetti | IT | 2014 | | 1/8 | | | ■ | | | | | | 0 | | | | | | 4/7 | ■ | ■ | | ■ | | | ■ | 5/49 | 10% |
| 9 | Celle ipogee Melinc | IT | 2004 | | 1/8 | | | ■ | | | | | | 1/6 | | | | ■ | | 3/7 | ■ | | | | | ■ | ■ | 9/49 | 18% |
| B | **Beverage industry** | | | | | | | | | | | | | | | | | | | | | | | | | | | | |
| 1 | UC Davis Wine Bre | CA | 2013 | | 0 | | | | | | | | | 2/6 | | | ■ | ■ | | 1/7 | | | | | | | ■ | 12/49 | 24% |
| 2 | Cantina Pizzolato | IT | 2016 | | 1/8 | | | ■ | | | | | | 1/6 | | | ■ | | | 4/7 | ■ | ■ | | ■ | | ■ | | 12/49 | 24% |
| 3 | Cantina Antinori | IT | 2013 | | 0 | | | | | | | | | 1/6 | | | | | ■ | 4/7 | ■ | ■ | | | | ■ | ■ | 7/49 | 14% |
| 4 | Bodegas Portia | ES | 2010 | | 0 | | | | | | | | | 0 | | | | | | 4/7 | ■ | ■ | | ■ | | | ■ | 6/49 | 12% |
| 5 | Macallan distillery | UK | 2018 | | 0 | | | | | | | | | 4/6 | | ■ | ■ | ■ | ■ | 4/7 | ■ | ■ | | | | ■ | ■ | 13/49 | 27% |
| 6 | Dalmunach distiller | UK | 2015 | ■ | 2/8 | | ■ | ■ | | | | | | 1/6 | | | ■ | | | 4/7 | ■ | | ■ | | ■ | | ■ | 13/49 | 27% |
| | | | | 7% | 4% | 0% | 2% | 42% | 9% | 4% | 4% | 0% | 7% | 7% | 2% | 7% | 36% | 18% | 16% | 19% | 47% | 33% | 4% | 31% | 18% | 18% | 71% | | |

**Fig. 4.1** Extract of the case studies analysis

and the presence of scenic views as important factors in promoting winery tours to tourists. Accordingly, process innovation and site attractiveness are of the utmost importance. Warehouses for food and beverage storage particularly stresses energy efficiency as vital to their financial sustainability.

Overall, the success of these cases is not so much dependent on the adoption of a single solution, which are often not particularly innovative themselves, but rather on the interaction of various strategies that combine, strengthening each other, in a coherent system capable of enhancing design choices in relation to corporate communication. As a result, a well-designed factory derives from a project considers the three main dimensions of the landscape together.

Good design makes a product useful.[3]

Quoting the industrial designer Dieter Rams and applying it to the field of architecture: good design emphasizes the usefulness of a product and avoids everything that could detract from it. However, since a product must be used, it must satisfy not only functional but also psychological and aesthetic requirements. The designer Rams stresses on the usefulness of these latter element, which should not only be decorative additions, otherwise they should be avoided completely.

Six patterns (Fig. 4.2) will now be examined, each focusing on a theme that is common to several projects and to which more than one good practice contribute.

[3] The quote is from a brochure distributed in occasion of Vitsœ opening in Royal Leamington Spa (United Kingdom), 2017, to which the Author participated. "The power of good design" includes Dieter Rams' ten points of good design.

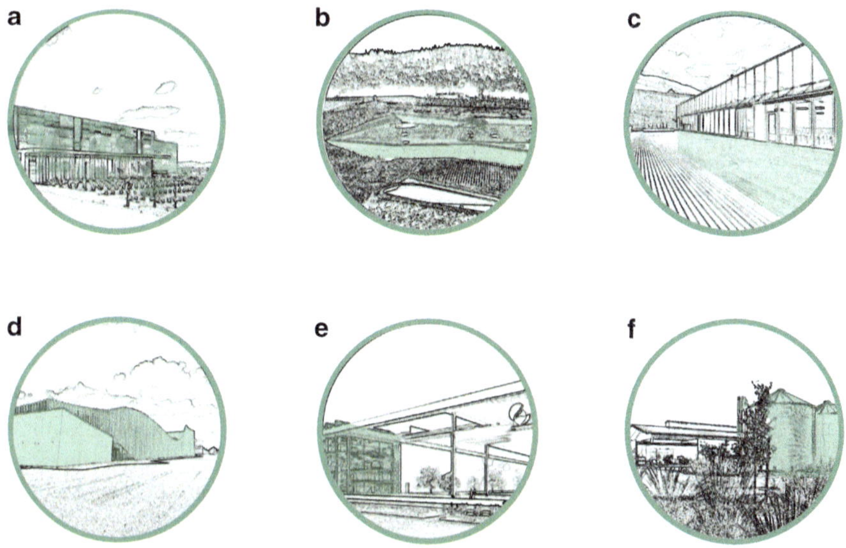


**Fig. 4.2** Six patterns from agri-food factories. Clockwise from top left: **a** Existing buildings and colour design; **b** Wine and territorial identity; **c** Food and underground buildings; **d** Dynamic perception from the road; **e** Multifunctional buildings; **f** Zero emission factories

**a. Existing buildings and colour design**. There are various methods to convey corporate image and respect for the landscape by means of architecture. Depending on the specific context, either contrast or consistency with the scenery can be acceptable. In fact, mimesis in not always advisable (Tandy, 1975), rather sometimes it can be useful to use singularisation or individualisation[4] (Hörmann, 2007). Regardless of the desired outcome, the colour design approach is often used in existing sites to hide or highlight certain features.

As a matter of fact, many constraints make retrofitting interventions more difficult and expensive in buildings that are already in place. Hence companies often prefer to intervene by reducing the visual impact of facilities first, through shielding screens (prefabricated or green), or colours and graphics.

Colour especially can have a significant impact on the built environment: in the industrial field it can be used to improve the perceived architectural quality of repetitive building types, such as a series of prefabricated structures, by modulating their appearance, harmonizing them with the surrounding environment, correcting their proportions, and deconstructing the visual mass at a low cost (Progetto Studio Colore, n.d.). There are several strategies for achieving these results, most of them based on practices used in art and design disciplines (Bell, 2004; Boeri et al., 2013; Spagnoli, 2010). From the use of sky tones that dematerialise the building on the horizon, to the harmonious division of the façade by means of tinted zones (deconstruction) or camouflage.

---

[4] That means defining controlled contrast with the surrounding to highlight the company identity while not detracting the landscape.

Even inside the factory colour can play an important role in improving the perception of space, safety, and orientation (Spagnoli, 2010). For instance, the feelings induced by different tones on workers can compensate for the sensations of heat and cold, or noise; the colour of machinery, paths, and signs can contribute to the reduction of human errors and risk factors, and can affect the well-being of workers too.

Although they belong to another industrial sector, the artworks by Ernst Von Garnier on the Thyssen Krupp factories represent a clear example of colour design and its potential: Rasselstein, Dortmund, and Stralsund are three big cities where his works can be found. However, as pleasant as the outcome is, attention must be paid to consider these examples sustainable factories, particularly in the way the superficial chromatism or graphic design of the façade is entrusted with the difficult task of mitigating massive and heavy industrial volumes in the scenery, and even when successful, the impact on the other landscape dimensions is neglected. So, colour design can help to enhance the visual quality of the building, but should be combined with environmental mitigation strategies, otherwise planning and design flaws will persist (Landscape Institute & IEMA, 2013; O.P.U.Q.P., 2015).

From the list, the graphic/colour approach is particularly evident in A3—Almazara Olisur (deconstruction and colour harmonization); B7—Carlsberg (use of light blue towards the sky).

**b. Wine and territorial identity**. Iconic architecture takes advantage of the bond between the territory and its typical products, an approach which has been registered in the beverage industry for many decades, especially in wineries and related facilities. This trend is supported by the scientific literature, which shows that medium and large-scale wineries have a vested interest in creating high-quality landscapes and implementing marketing strategies that strengthen the link between the landscape and its traditional products through the functional use of the scenery (Harea & Eplényi, 2017; Tassinari et al., 2013; Torquati et al., 2015). This scenario goes far beyond the typical division between production and consumption and opens up new spaces for food and beverage tourism (Qualizza, 2012).

Hence B3—Cantina Antinori, B4—Bodegas Portia, B5—Macallan distillery, and B6—Dalmunach distillery, to name a few from the list, planned their factories as environmentally friendly buildings with spectacular design, where visitors are invited to take a tour which frequently concludes in dégustation areas, restaurants, or shops. The factory becomes a tourist attraction as well as a sign of quality in the area: a territorial landmark.

Even though some of these wineries and distilleries have sizes and needs smaller than larger industrial complexes, the subject might still be informative to investigate in regard to the success they have gained. Both have a close connection to their surroundings and to their environmental and aesthetic character since they involve nature in the whole value chain. As a result, the finest strategies and tactics that small-size beverage industries have developed appear to be easily transferable to larger food factories.

**c. Food and underground buildings**. Landform design and hypogeous warehouses are often used to reduce energy demand, mitigate the visual impact of the facility, and save money. Several of the analysed food companies show that they have followed in the footsteps of the beverage industry in understanding the communicative potential of a sustainable factory in tune with the landscape, something which is consistent with company values of naturalness and environmental health, as well as the identity of the territory.

Thus, in the food and beverage industry, it is frequent to observe underground or semi-underground buildings, where the visual mitigation of its impact on its surroundings—which is often characterised by cultivated land—is combined with the thermoregulatory properties of the ground. This in fact can provide excellent conditions for the storage and preservation of food or, at least, reduce the amount of energy required for cooling the warehouses (Mazarrón et al., 2012).

As a result, some interesting examples as A1—Melinda, A2—Bio Pastificio IRIS, A3—Almazara Olisur, and A9—Celle ipogee Melinda, have developed complexes in which environmental, passive, and active strategies are integrated with aesthetic and social ones.

**d. Dynamic perception from the road**. Precise lines, graphs, lights, and symbols offer the car driver a brief but memorable vision of the factory. We might call it the "wow factor". When a plant is located near a major road or highway the potential for communication is extraordinary. The road is the stage on which the companies present themselves: through a strong and identifiable sign in the landscape as seen by a fast-moving car, where the rapid and effective perception of flashes of colour, familiar symbols, and icons—or rather "memorable" vision—plays a crucial role in promoting the company and communicating its ethos. The kinetic experience of spaces is thus managed according to the driver's point of view (Cipriani, 2012), in a play of space and movement, of lights and textures, symbols and shapes that follow on from one another (Appleyard et al., 1964).

In this context, projects range from "pioneering forms of highway colonisation for self-promotional purposes", which is an expression of an individual business, to more recent "forms of linear city on a territorial scale", which are configured as large industrial polygons or scientific-technological parks (Vitali, 2014).

This pattern includes many projects from the list , such as A4—Stabilimento Maina, A5—Grandi Salumifici Italiani, A8—Pastificio Felicetti, A23—Carnitalia, B7—Carlsberg.

**e. Multifunctional buildings**. The production process is put side by side with social and sport facilities such as parks, playgrounds, a museum, a restaurant or a shop, spaces for public events, skate parks, and climbing walls. Drawing directly from energy infrastructure where it is defined as "integrated multi functionality" (Ginelli & Daglio, 2016), this pattern attempts to put together places of production and everyday life by making the former more liveable and attractive to the local community. The goal is to reintroduce manufacturing areas in the city by engaging various groups of inhabitants in social, leisure, cultural, and educational events that also demonstrate the site's environmental quality.

This approach adds value to the company, which allows the community to utilise the surrounding places for diverse uses as part of the city's overall enrichment, while enhancing its role and social intent (Hörmann, 2007). Some exponents from the list are: A18—Iroha Village Factory, A21—Van Gelder, and B13—St. James Gate Brewery.

**f. Zero-Emission factories**. The last pattern that has emerged in the collection of projects is the showcasing of good environmental practices and technical equipment as formal and educational means. This category includes environmentally sustainable factories, zero-emission factories, environmentally fair factories, and passive factories, which still appear to be exceptions to the rule (Dombrowski & Riechel, 2013). Luckily, they have achieved widespread media attention and are now ideal models for many businesses.

They often display as formal design elements the devices and technologies employed to achieve the excellent environmental performances that distinguish them. These features appear to evoke ideas of essentiality and usefulness, sometimes whilst also pursuing didactic goals. They convey the idea that communicating to be green is as important as being green itself.

Some use photovoltaic façades to both reach energy self-sufficiency and communicate their green performances. Others use water collection tanks, grain-storage silos, lamination basins, phytoremediation areas for waste waters not only as functional elements but also as a way to feature the visual quality of the site.

Excellent examples from the list are: A2—Bio Pastificio IRIS, A6—olare Manufaktur Peter, and B1—UC Davis Wine Brewery.

To summarise, this chapter discussed how respect for the environment, local communities and territorial quality by companies is growing. In particular, the image and architecture of the factory appears increasingly important for competitiveness, and the evocative design quality that is traditionally reserved for headquarters or flagship stores is progressively being extended to production plants in the food and beverage industry. Actually, the many examples of corporate identity projects entering architectural magazines from other sectors—mostly automotive and textile industries—show the increasing relevance of the topic for a great variety of functions (The Plan Art & Architecture, 2011). However, at times in these cases there is a risk of falling into what Qualizza (2012) defines as the "Disneyfication of the work space", when the factory becomes more and more a mere narrative device, a theatre in which to stage the spectacle of the production of goods or services, rather than a place for producing and working.

Food and beverage industries, perhaps facilitated by their proximity with nature, are instead demonstrating that valuable opportunities for both the company and the territory are found when good environmental performances and perceptive and social qualities in factories are combined. Thus, the successful strategies and practices implemented by the sector have a strong up and out scaling potential. Now, it is time to learn how to put this lesson into practice.

## References

Appleyard, D., Lynch, K., & Myer, J. (1964). The view from the road. *MIT*. https://doi.org/10.2307/142014.

Bell, S. (2004). *Elements of visual design in the landscape* (2nd ed.). Spon Press.

Boeri, C., Braz de Oliveira, I., & Giambruno, M. C. (2013). Colour design and industrial landscape. *III CITCEM Conference*.

Bortolini, M., Gamberi, M., Pilati, F., Regattieri, A., & Accorsi, R. (2017). Design of an Innovative Plant for the Wastewater Recovery and Purification in the Food & Beverage Industry. In G. Campana (Ed.), *Sustainable Design and Manufacturing 2017* (Vol. 68, pp. 691–700). Springer. https://doi.org/10.1007/978-3-319-57078-5.

Cipriani, L. (2012). *Vita e morte del paesaggio industriale. Indagini e proposte per la marca trevigiana in trasformazione*. Università degli Studi di Trento.

Clark, J. P. (2005). Food plant design. Food engineering. In G. . Barbosa-Canovas (Ed.), *Enciclopedia of Life Support Systems* (pp. 683–696). EOLSS Publisher/UNESCO.

Dombrowski, U., & Riechel, C. (2013). Sustainable factory profile: a concept to support the design of future sustainable industries. *11th Global Conference on Sustainable Manufacturing*, 72–77.

Enea. (2016). *Rapporto Annuale Efficienza energetica (RAEE) 2016*. https://doi.org/10.1007/s13398-014-0173-7.2.

Fanfani, R., & Pieri, R. (a cura di). (2016). *Il sistema agro-alimentare dell'Emilia Romagna. Rapporto 2015* (Unione regionale delle Camere di commercio dell'Emilia-Romagna Regione Emilia-Romagna; & Assessorato agricoltura caccia e pesca (eds.)). e-book.

Fedriga, P. (2016). Una crescita sostenibile nel futuro del Pianeta. *Agricoltura*, 24–26.

Gandolfi, C. (Ed.). (2011). *Paesaggio e Attrattività*.

Ginelli, E., & Daglio, L. (2016). Le infrastrutture per le energie rinnovabili nel paesaggio. Strumenti di progetto e traiettorie dell'innovazione. *TECHNE*, *11*, 119–126. https://doi.org/10.13128/Techne-18410.

Giovanardi, M., & Lucarelli, A. (2018). Sailing through marketing: A critical assessment of spatiality in marketing literature. *Journal of Business Research, 82*(2018), 149–159. https://doi.org/10.1016/j.jbusres.2017.09.029.

Harea, O., & Eplényi, A. (2017). Viticultural Landscape Patterns – Embedding contemporary wineries into the landscape site. *Landscape Architecture and Art*, *10*(10), 7–14. https://doi.org/10.22616/j.landarchart.2017.10.01.

Hörmann. (2007). Industrial Architecture. *Portal*, *11*.

Industria alimentare. (2012). In *Enciclopedia Treccani*. https://www.treccani.it/enciclopedia/industria-alimentare_%28Dizionario-di-Economia-e-Finanza%29/.

Landscape Institute, & IEMA. (2013). *Guidelines for Landscape and Visual Impact Assessment (GLVIA)* (Landscape Institute and Institute of Environmental Management & Assessment (Ed.); 3rd ed.). Routledge. https://doi.org/10.1057/palgrave.udi.9000084.

Lee, W., & Okos, M. R. (2011). Sustainable food processing systems—path to a zero discharge: Reduction of water, waste and energy. *Procedia Food Science, 1*, 1768–1777. https://doi.org/10.1016/j.profoo.2011.09.260.

Marchi, L. (2020). Towards the Harmonization of Industrial Facilities with the Landscape. A catalogue of good practices. *SMC - Sustainable Mediterranean Construction, 12*, 156–160.

Mazarrón, F. R., Cid-Falceto, J., & Cañas, I. (2012). Ground thermal inertia for energy efficient building design: A case study on food industry. *Energies*, *5*(2), 227–242. https://doi.org/10.3390/en5020227.

O.P.U.Q.P. (2015). *Linee guida per i paesaggi industriali in Sardegna*.

Philipp, M., Schumm, G., Peesel, R. H., Walmsley, T. G., Atkins, M. J., Schlosser, F., & Hesselbach, J. (2017). Optimal energy supply structures for industrial food processing sites in different countries considering energy transitions. *Energy*, 1–12. https://doi.org/10.1016/j.energy.2017.05.062.

Progetto studio colore. (n.d.). Retrieved August 26, 2019, from http://www.progetticolore.com/it/ambienti-progetti-studio-colore/ambiti-di-intervento?-ida=4.

Qualizza, G. (2012). Inventare lo spazio di una green factory. In M. Cavallo, P. Degli Esposti, & K. Konstantinou (Eds.), *Green marketing per le aree industriali. Metodologie, strumenti e pratiche* (pp. 220–238). Franco Angeli.

Regione Lombardia. (2011). *Paesaggi periurbani. Linee guida paesaggistiche per il governo del territorio / Periurban landscapes. Landscape planning guidelines.* http://territorio.regione.emilia-romagna.it/paesaggio/pubblicazioni/paesaggi-periurbani.

Spagnoli, M. (2010). Friedrich Ernst von Garnier: un pioniere del color design Friedrich / Ernst v. Garnier: a true color design pioneer. In IDC Colour Centre (Ed.), *(a cura di C. Boeri),Colore. Quaderni di cultura e progetto del colore.*

Tandy, C. (1975). *Landscape of industry.* Leonard Hill Books.

Tassinari, P., Torreggiani, D., Benni, S., & Dall'Ara, E. (2013). Landscape Quality in Farmyard Design: An Approach for Italian Wine Farms. *Landscape Research, 38*(6), 729–749. https://doi.org/10.1080/01426397.2012.746653.

The Plan Art & Architecture (Ed.). (2011). *Corporate Identity. Architecture in detail.* Scripta Maneant.

Torquati, B., Giacchè, G., & Venanzi, S. (2015). Economic analysis of the traditional cultural vineyard landscapes in Italy. *Journal of Rural Studies, 39*, 122–132. https://doi.org/10.1016/j.jrurstud.2015.03.013.

Vitali, P. (2014). I luoghi della produzione nei territori della deindustrializzazione. Il caso Bergamo. In S. Marini & F. De Matteis (Eds.), *La città della postproduzione* (Edizioni N, pp. 61–71). https://doi.org/10.4458/1921.

# Chapter 5
# A Toolkit to Assess and Mitigate the Impact

**Abstract** This chapter introduces a tool that may assist firms in recognising and mitigating their negative effects on the landscape. It consists of an Assessment System and a Catalogue of good practices. With the aim of overcoming the gaps in already available tools, the System adopts a multi-criteria approach and modifies the U.S. GBC LEED protocol by including aesthetic-perceptual and social considerations in addition to environmental concerns. The Catalogue is a collection of exemplary factories that were chosen for their adoption of good design practices across multiple landscape dimensions. In turn, a library of general mitigation tactics is compiled, organised by subject, and linked to specific impacts as assessed by the System. This multifaceted classification allows for the simpler retrieval of tactics which are connected by the issues they mitigate. The subsequent and iterative application of the two complimentary tools forms the 'protocol' for designing sustainable factories, applicable to both new constructions and renovation projects. The chapter ends by outlining the toolkit's general application, in particular targeting businesses and designers.

**Keywords** Best practices · LEED · Mitigation tactics · Multi-criteria assessment · Perceptual impact

Despite encouraging progress, as illustrated in previous chapters, a huge gap still exists between theory and practice. On the one hand, academics and policymakers demand a more comprehensive approach to the subject, addressing all the various facets and scales involved in the relationship between industrial facilities and the landscape. By proving that many benefits and opportunities are provided through this process even when no incentives are offered, companies are encouraged to change their mindset. On the other hand, because sustainability is still largely regarded by companies as optional, economic restrictions and a lack of expertise prevents companies from building new or renovating existing factories in harmony with their surroundings. In some cases, even when a firm is informed of the associated benefits and decides to limit its negative impact whilst improving the image it presents to its consumers, there is the risk of green or social-washing if interventions are sectorial and trivial.



L. Marchi, *Designing Sustainable Factories*, Advances in Global Change Research 72,
https://doi.org/10.1007/978-3-030-92227-6_5

To this end, Chap. 3 has illustrated that some tools have been developed to encourage truly green, environmentally friendly factory design. BREEAM Industrial, DGNB Industrial Site, and Protocollo ITACA Edifici Industriali are some clear examples. Instead, only a few instruments with limited scope have been devised to reduce the aesthetic and perceptual effects of factories on the landscape. This sectorial approach to a factory's landscape compatibility has been discussed to have considerable limits in regard to the exploitation of uncaptured values and synergies, and to be ineffective in guiding companies. Instead, pragmatic, comprehensive and easy to apply tools would be more suitable, and would also be capable of acting on the leverage points of the system. Moreover, as presented in Chap. 1, the heterogeneity and complexity of industry require tools and actions that are highly adaptable on a case-by-case basis, while many of the available instruments are place-specific.

On these premises, this chapter describes a research project aimed at assisting companies in the design of more sustainable factories. The study has been conducted by the Author from 2016 to 2019, during her Doctorate in Architecture at the University of Bologna (Italy). The project was funded by the Italian food company Società Cooperativa Agricola Orogel S.p.a. and was expected to identify mitigation strategies for integrating industrial complexes into the landscape.[1] Following on from this, the Author was awarded a postdoctoral grant from the University of Bologna to continue her research, from 2020 to 2022.

**Aims and objectives**. This study seeks to contribute to the development of measures to mitigate the effects of manufacturing plants on the landscape. This will be accomplished through a multidisciplinary approach that combines, in a single procedure, several intertwining themes: from the interferences of the facility with the aesthetic-perceptual dimension of the landscape, to damage caused to the physical environment that hosts the building; from disturbances affecting the social component of a territory, to the functional needs of manufacturing, as well as the communication requirements for the promotion of the company in order to boost its competitiveness (Fig. 5.1).

The review of scientific literature and the observation of real-world practices from previous chapters have informed this procedure so that it has been designed for the direct use of enterprises that aim to design their factories to be environmentally and socially friendly and, at the same time, to use the quality of their site as a source of market growth.

Using the words by Giancarlo De Carlo—a famous Italian architect from the twentieth century—the aim is to

> deconstruct an imaginary stuck on known reality by offering an overview of concrete and non-standard possibilities (De Carlo, 2013, p. 14).[2]

The idea that the design of factories should only respond to functional and profit criteria is neither acceptable nor beneficial to open-minded and far-sighted

[1] The specific task as in the contract was "Study of the relationship between industrial site and the context though strategies of landscape integration, marketing and promotion of food culture".

[2] The original Italian text is: "L'obiettivo è scardinare un immaginario bloccato sulla realtà nota offrendo una panoramica di possibilità concrete e non standardizzate".

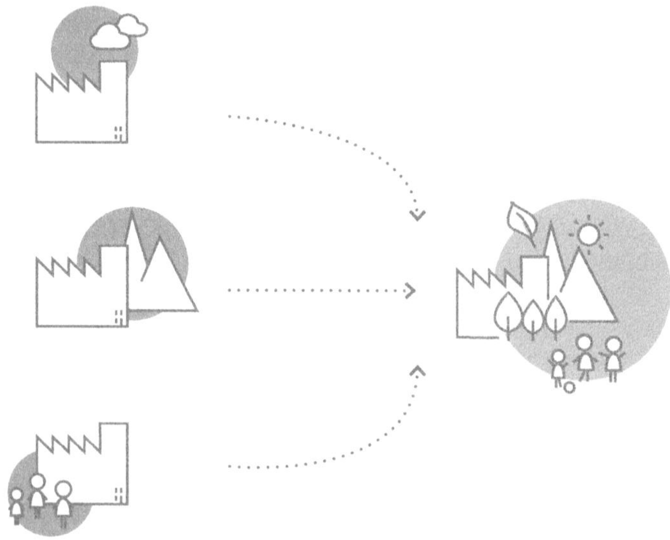

**Fig. 5.1** This study combines environmental, aesthetic-perceptual, and social aspects in one tool

entrepreneurs. Therefore, this study supports companies and designers by delivering a structured method for the analysis of the project and to offer them the means to enhance its quality. The study's scope is to help companies define their project through the implementation of best available practices, according to each firm's functional requirements, spending capacity, aesthetical preferences, and values. This translates to a tool that can manage complexity, rather than having to rely on models or pre-designed solutions. Thus, the goal is to provide businesses with a design support tool which can analyse the problem and identify case-by-case adaptable strategies to mitigate negative effects while boosting positive ones.

The main novelty of this research lies in its attempt to combine different aspects that have been considered separately up till now. Thus, the three landscape dimensions are integrated into a coherent framework, as well as the three interconnected scales (i.e., site, building, and process). Moreover, this study focuses on both the design of new industrial sites and the renovation of existing facilities, where the latter are often neglected by available tools.

**Methodology**. The methodology used to develop a toolkit for designing landscape-compatible factories is based on the enhancement of existing assessment tools and systematising best practices that have already been implemented in real-world situations. In order to define the toolkit, the process has been broken down into three main steps:

1. To identify a comprehensive list of interferences on the landscape by industrial facilities. Then, to select suitable indicators to measure individual impacts and their weighted effects. This is followed by the setting of a coherent system of credits to measure the effects/problems and prioritise interventions.

2. To collect exemplary case studies as references for the mitigation of detrimental effects. This provides a catalogue of projects and related good practices, from which an inventory of relevant mitigation tactics can be extracted.
3. To combine the two instruments into a consistent procedure (also referred to as the 'protocol') that companies and designers can adopt for the design of more sustainable factories.

Further details on the general methodology can be found in Marchi et al. (2020). In a nutshell, the outcome of this study consists of two complementary tools: an impact Assessment System and a Catalogue of good practices. These are applicable either to new construction or renovations adopting a consistent approach that varies the specific requirements when needed.

**Limits**. The toolkit has been specifically designed for agri-food companies since the peculiarities of the sector make them particularly suitable (as discussed in Chap. 4). In particular, the practices within the Catalogue have been retrieved from exemplary food and beverage companies. As a result these tactics are custom-made for this industry. However, most of them are relevant and easily adaptable to other industries and types of buildings too. Indeed, the impacts and performances that the Assessment System is meant to measure are generally applicable for other sectors: when sector-specific requirements are incorporated, they are identified by ad-hoc symbols in the info-sheets.

In terms of geographic coverage, the case studies, practices, and tactics come from all over the world, even if Europe—and especially the Mediterranean region, where the food and beverage sector is still a major industry—has a greater sensitivity to the issue, as evidenced by the geographical balance of cases included in **Annex 1**. In addition, a detailed investigation of the Emilia-Romagna region has been completed as part of the Orogel case study, because this needed having a clear understanding of the environmental, socio-cultural, and economic systems to which the project relates. Nevertheless, the toolkit is devised to be used in many other contexts provided that local norms and regulations are observed.

## An Assessment System to Detect and Measure the Impact

A lack of knowledge and awareness about the negative impact of factories on the landscape, as well as related missed opportunities and benefits, have been shown to inhibit the design of sustainable sites. This is without mentioning the possible synergies that might be discovered through a well-structured analysis of the project. For these reasons, the first step is to assist businesses in better understanding the impact of their facilities, whether these are already operating, or are new constructions at the design stage. This entails evaluating the negative effects on the environment, the scenery, the buildings' occupants, and the local community, and then weighing each interference against the overall performance of the building and considering the site and the unique features of the company.

A systematic review of the scientific and trade literature carried out by the Author has highlighted that recurrent interferences of factories on the landscape largely vary in the features and indicators that are used. On the one side, most effects on the environment are quantifiable and their measurements make use of well-established methods and shared parameters (e.g., LCA indicators, energy performance indexes, ratios of physical measurements). This category addresses for instance the soil permeability at site level (e.g., percentage of hardscapes), the energy demand of the building (e.g., energy for heating per year per floor unit), and the indoor lighting levels (e.g., lumen per square metre). On the other side, perceptual and social disturbances are more difficult to measure due to the qualitative nature of human perception, which makes it difficult for it to be measured by quick, objective, and shared methodologies (Tolli et al., 2016). This second category addresses, for example, the visual quality of the factory, its harmonisation with the scenery, and the willingness of the local community to see the building being constructed next to their homes. In fact, these often require field surveys and involve both the critical judgement of experts and the preferences of communities (Butler & Berglund, 2014; Cassatella & Peano, 2011; Rosley et al., 2013; Tudor, 2014). Currently, only a few methods and metrics are available for quantifying subjective interferences of a construction on the landscape, even in procedures that require it like the Environmental Impact Assessment (Sowińska-Świerkosz, 2016; Tolli et al., 2016).

Therefore, since many different aspects and scales are involved, this study assumed the multi-criteria structure of Green Building Rating Systems as capable of managing impacts of varying natures and synergies.[3] The effectiveness of these tools in communicating complex performances to the market via basic and easy-to-understand outcomes was also a part of the decision.

After analysing GBRSs applicable to industrial facilities, the U.S. GBC LEED was selected as the most suitable for the purposes of this research. In particular, the scheme *LEED v4 for Building Design and Construction: Warehouses and Distribution Centres* was chosen because of the highest number of credits corresponding to factories' recurrent impacts, because it is one of the most widely spreaded worldwide (Doan et al., 2017), and, not least importantly, because it can be easily modified and integrated, even with new assessment categories (U.S. GBC, 2013). Further details about this stage of the methodology can be found in (Marchi & Antonini, 2021).

Despite appearing the most suitable, the available LEED scheme lacks a comprehensive assessment of perceptual-aesthetic as do all the other GBRSs. Thus, a new category of credits addressing the perceptual-aesthetic impact was conceived. These were retrieved or adapted from specific studies and guidelines, among which

[3] The functioning of GBRSs is based on the evaluation of a vast range of green buildings' performances (such as energy consumption, water use, indoor environmental quality, location) by means of specific baseline. Thus, the measures of these performances are weighted, using a balancing process specific to each scheme, and are combined into a single grade that shortly communicate the building's overall level of sustainability. Using the LEED nomenclature, within each "category" (e.g., Energy and Atmosphere), specific performances are associated with "credits" (e.g., Renewable energy production) that can give at maximum 3 points out of 110. The sum of points achieved in all selected credits gives the final certification level. That is, Certified (40–49 points), Silver (50–59), Gold (60–79) and Platinum (>80).

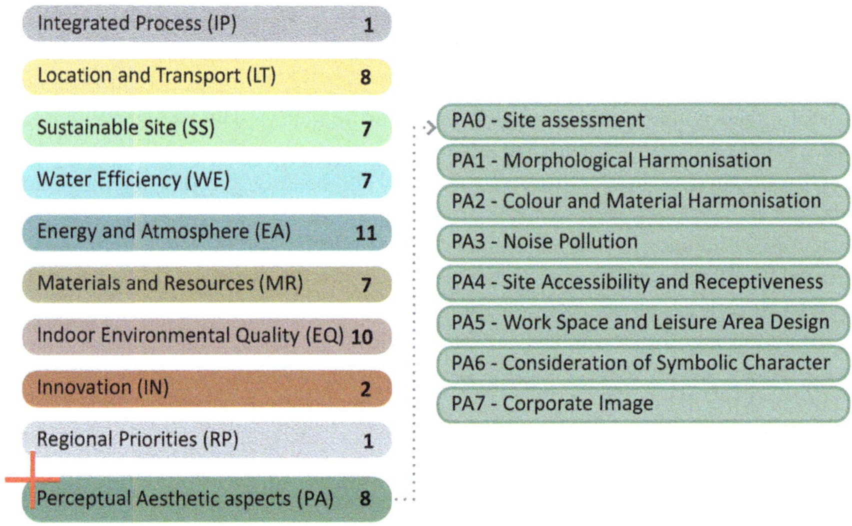


**Fig. 5.2** On the left, the assessment system evaluates the project via 10 different categories: nine are from LEED, while the last is newly added. On the right, the eight new credits within the PA category

were those mentioned in Chap. 3. Figure 5.2 illustrates the eight newly proposed credits, namely: 1. Site assessment; 2. Morphological Harmonisation; 3. Colour and Material Harmonisation; 4. Noise Pollution; 5. Site Accessibility and Receptiveness; 6. Workspace and Leisure Area Design; 7. Consideration of Symbolic Character; 8. Corporate Image.

The first credit, PA 0 "Site evaluation", is a prerequisite because the user cannot proceed with the assessment of the intervention without a thorough understanding of the character and features of the context. Contextualisation is essential in sustainable design, as the design team must be able to find links and synergies between the project and its broader system in order to determine optimal actions based on the cost–benefit ratio in the short and long-term (U.S. GBC, 2014). Credits from 2 to 4 assess the project's physical consistency with respect to the aesthetic-perceptive sub-components of the landscape outlined by the Landscape Character Assessment (Tudor, 2014): e.g., shape, layout, colour, materials, and patterns. Whereas credits from 5 to 8 focus on the project's socio-cultural landscape aspects: e.g., how it helps the company to convey its performances and values; how it attracts residents and visitors; how comfortable their employees are working there.

For the quantification of the LEED "original" credits, direct reference is made to the LEED User Manual (U.S. GBC, 2017). While for the new PA credits, new info-sheets have been defined: these have the same structure as LEED but differ in terms of procedures, tools, and indicators, as they are primarily aimed at assessing qualitative impacts.

Each card (Fig. 5.3) contains information and appropriate bibliographic references that the user/evaluator will need to assign a score to a single credit. Specifically, the card includes:

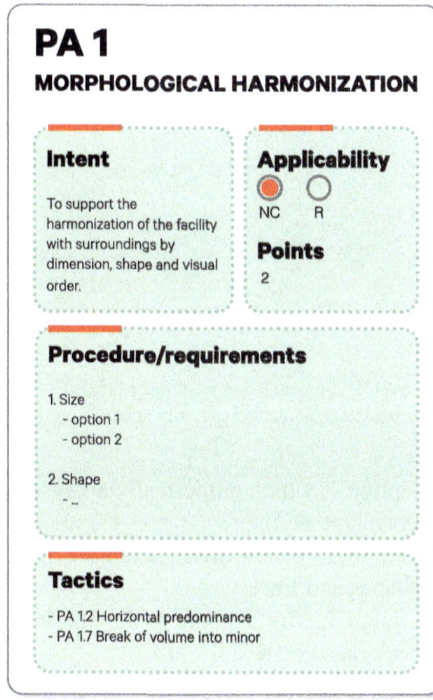


**Fig. 5.3** Example of card for credits

- the intent, which illustrates the purpose of the credit;
- the applicability of the credit to new buildings (NC) and/or renovation of existing constructions (R);
- the maximum points attainable;
- the procedure, indicators, and investigation tools that can be used to assess the credit or one of its sections;
- the general tactics that can be applied to the project to mitigate the impact, which is a link with the Catalogue of good practices.

In the eight new cards at least one indicator per credit is associated. This can be quantitative, qualitative, or semi-quantitative, depending on the nature of the specific impact. Given that there is no single method for measuring the degree of interference of a building with respect to the aesthetic-perceptive and socio-cultural dimensions of the landscape, the choice of indicators and measurement methods may vary on a case-by-case basis. Therefore, in some credits different methods are provided, namely "options". These can be chosen based on the information available, the goal of the assessment, or specific technical/time restrictions. For example, the credit PA1 "Morphological Harmonisation" relates the sub-components (i) Size, (ii) Shape, and (iii) Visual order of the scenery with the specific features of the project and offers the user several options for assigning the score. Concerning Shape, the relationship can be assessed by means of:

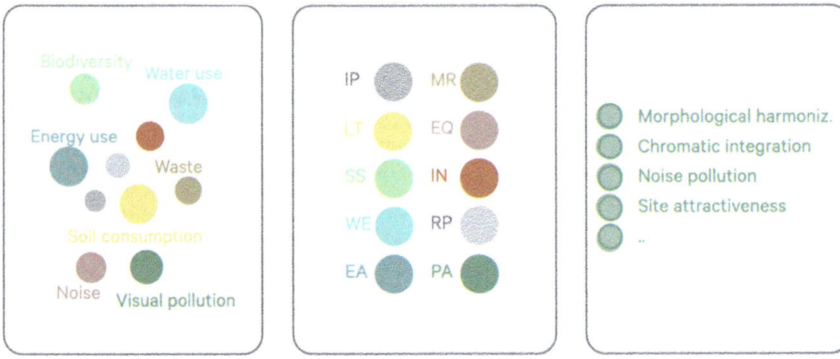


**Fig. 5.4** The assessment system in brief. From left to right: recurrent impacts of factories (i) are organised and measured (ii), and specific cards are developed for the new credits (iii)

- Option 1, which numerically analyses the shape ratio, the footprint ratio, and the prevalent skyline;
- Option 2, which offers a table to fill according to the perceptive judgment of shapes and lines;
- Option 3, which recurs to the Shape Disharmony Index (SDHI) from Sowińska-Świerkosz (2016).

As a result, the System's version for new constructions is composed of 10 Credit Categories with 12 prerequisites and 50 credits (Fig. 5.4), whilst the version for retrofitting existing buildings is made up of 42 credits, as some choices such as the location and the features of the bearing structure are fixed.

As for the original LEED scheme, the Assessment System awards factories based on their level of sustainability: from Certified to Platinum. The difference is in the maximum number of points achievable, which in the revised version for new constructions is 127 (compared to 110 of U.S. GBC).

In truth, adding an entire new category in LEED would require a series of complex steps in order to gain official approval. However, for the scope of this study it has been assumed that the new PA category had already received the GBC agreement.[4]

## A Catalogue of Good Practices to Inspire Mitigation Action

As firms are often unaware of what they can do to lessen their impact, the second research activity aims at providing them with a collection of excellent examples and

[4] In concrete terms, before becoming permanent in LEED, the eight new credits should undergo the process of being Pilot Credits, which are credits selectable from an online library at U.S. GBC website (U.S. GBC, n.d.) and then included in the Innovation category (IN). At present the Innovation category allows for a maximum of five points to be achieved, thus only a couple of the new credits could be chosen.

practices which might inspire the design process. Since a review of the literature has pointed out that manufacturers require tactics that are simple to grasp and implement, this approach is both pragmatic and hands-on.

The Catalogue which results from this stage is composed of two main elements: case studies, and tactics. The first serves as a concrete reference in which specific practices[5] from real cases are illustrated, the second as general guidance for topics common to two or more case studies. In other words, tactics[6] are the starting point for adapting a general approach to a specific practice based on the project features and requirements.

In accordance with the multi-criteria structure of the System, the collection comprises a wide range of subjects affiliated with green buildings and landscape design that are classified by categories retrieved from the preceding tool (e.g., Location and Transportation; Sustainable Site, Perceptual-aesthetic aspects). Hence, tactics and practices can be selected and referred to on a case-by-case basis.

As briefly mentioned in Chap. 4, the process of building the catalogue started with the selection of exemplary agri-food facilities that implemented good practices from at least one of the following areas:

- reduced impact on the environment at site, building, and/or process level;
- mitigated perceptual-aesthetic disturbances with the scenery;
- protection of the well-being and health of occupants and neighbours.

The cases are retrieved through a keywords search from trade literature including corporate websites and reports, international design competitions and awards,[7] architectural magazines, and online databases, but also from guidelines and studies mentioned in the previous chapters. Then two subsequent and increasingly in-depth types of examinations are performed—namely "horizontal" and "vertical" analysis. Further details on the procedure can be found in Marchi (2020).

The horizontal analysis comes first, and helps to determine whether the case is appropriate for the purpose and deserves further investigation. In short, it determines how many System credits the project uses. If more than two credits are involved, the project is accepted and moves on to the next stage.

---

[5] Practice means a specific design choice aimed at satisfying one or more building mitigation requirements through the technological, formal, or aesthetic attributes of a physical part of the building (e.g., external thermal insulation system, biomass cogeneration plant, expansion in continuity with the existing complex, colour division of the façade), a functional structure (e.g., relax room for workers, equipped green area) or a management activity (e.g., visits to the production process).

[6] A tactic is a non-specific design approach to satisfy one or more building mitigation requirements through the technological, formal, or aesthetic characteristics of a physical part of the building (e.g., thermal insulation, cogeneration plant, expansion of the building, color design), a functional structure (e.g., services, relational spaces) or a management activity (e.g., events, promotional activities).

[7] For example: Green Company Award—Industria e Paesaggio (http://www.abitare.it/it/news/2016/10/08/green-company-award-2016/?refresh_ce-cp, accessed on 12 March 2018) or Brand&Landscape Award (https://concorsiawn.it/brand-landscape/home (accessed on 15 June 2017).

The vertical analysis aims at developing the case studies' cards and identifying good practices. The cards, or info-sheets, are made of two pages where data is organised into six sections, as illustrated in Fig. 5.5. The first page (cover) provides general information on the project, such as the function, geographical and time references, and the type of intervention that indicates whether it is a new construction (NC) or a retrofitting of an existing building (R). The second page describes in detail the good practices for which the project has been selected. The summary in Sect. 4 helps the reader to choose whether the case is worth investigating further depending on their needs. The bulleted list in Sect. 5 describes the good practices within their context and their cross-links. Section 6 acts as navigator for the Catalogue, summarising the good practices using keywords and referring them to the credits of the System, hence leading to the extraction of the tactics.

From the complete list of practices—which are case-specific—general tactics are extracted according to two alternative rules: when two or more practices from the same credits have a common denominator, or when the same practice recurs in two or more cases. As a third rule, tactics can also be drawn directly from literature sources, such as guidelines or topic-specific research projects. An example of the first rule might help the reader to better understand the process. Within the credit PA1 "Morphological harmonisation", both practices "green hill" and "land depression" operate on the landform: the former to hide the height of the building; the latter to

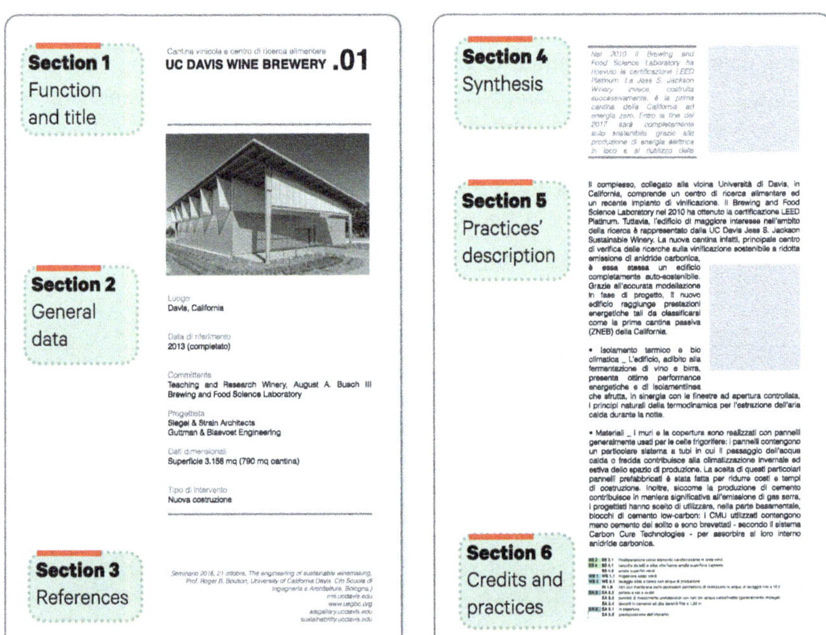


**Fig. 5.5** Example of a case study card. The cover (left) provides general facts about the project, whilst the second page (right) illustrates the good practices in detail

**Fig. 5.6** Example of a tactic card

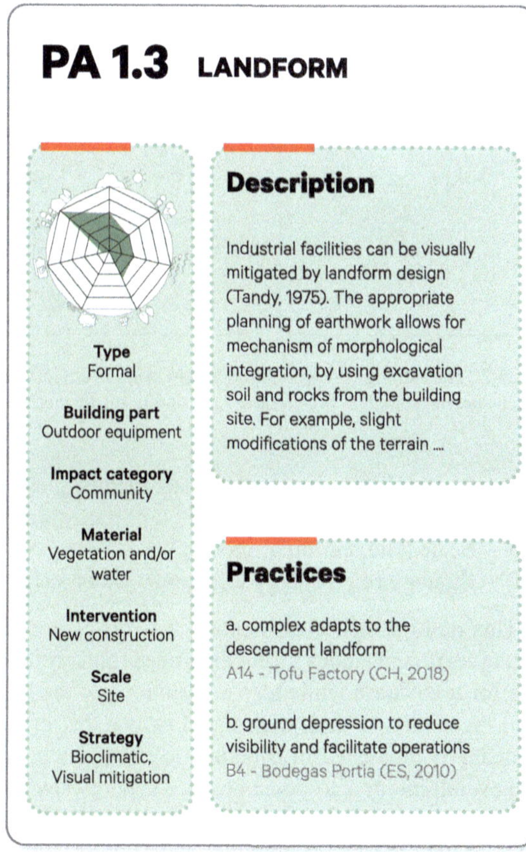


reduce the visual exposure of the complex. As a result, "PA1.3—Landform" is taken as a tactic. The alphanumerical code preceding the tactic's title is the same as that of the credit from which the tactic belongs (a consistent colour code has also been applied to facilitate the connection).

Each tactic is then described in a card that contains (Fig. 5.6): the general description (what it is; in what ways helps; how it applies; bibliographical references), and the bulleted list of practices from which it derives, along with the related case studies. In addition, the info-sheet includes a classification of the tactics via seven facets, intended to facilitate the user leafing through the contents of the Catalogue.

The categories are based on those frequently used by architectural search engines.[8] They are:

(i) Type (e.g., technological; formal; functional)
(ii) Building part (e.g., load-bearing structure; vertical envelope)
(iii) Impact category (e.g., climate change; health and well-being; biodiversity)

[8] For example, The Plan, ArchINFORM, ArchDaily, Divisare.

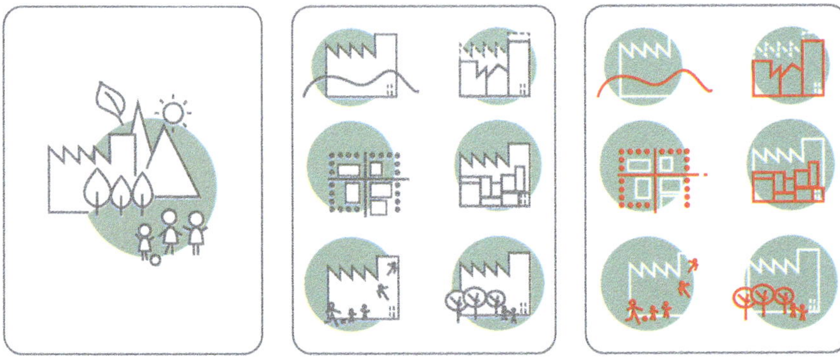

**Fig. 5.7** The catalogue of good practices in brief. From left to right: case study info-sheets (i) from which specific practices are retrieved (ii) and in turn general tactics are extracted (iii)

(iv) Material (e.g., wood; concrete; metals; glass; vegetation)
(v) Intervention (new construction or renovation)
(vi) Scale (site, building, process)
(vii) Strategy (e.g., energy efficiency; water saving; visual mitigation).

This multifaceted classification was developed for several purposes. Firstly, to help users to find tactics fitting their needs and preferences (e.g., seeking a tactic suitable for renovation while also contributing to the broader strategy of visual mitigation). Secondly, to find interrelated tactics that are mutually beneficial (e.g., if the company is interested in Bioclimatic strategies, it will look for all related tactics). The overall intent is to boost positive synergies while optimising design choices, as well as construction activities. For instance, "PA 1.3—Landform" can relate to "EA 2.5—Ground thermal properties" and "PA 3.1—Noise screen by landform" since a green hill can help to visually mitigate the building, take advantage of the thermal inertia of the ground, and reduce the noise from/towards its surroundings.

As a result, currently the Catalogue consists of 45 case studies from which 295 good practices have been selected, and 106 general tactics devised (Fig. 5.7). The collection of tactics is included in **Annex 2**. Every component of this tool is open to future integration (cases, practices, or tactics), making the structure flexible for progressive additions.

## A Toolkit to Design Sustainable Factories

The two tools presented above constitute the Toolkit, which is intended to be used directly by firms. When the company does not have an house qualified professional, the task of implementing it could be assigned to an outside person/team such as an architect, engineer, landscape architect, or another skilled professional. This person/team will also be responsible for involving stakeholders when needed and in the most effective way possible. Thus, the toolkit's main target groups are companies and practitioners.

The procedure to follow for the application is referred to as "the protocol". It requires a first evaluation phase that provides information about the case study's critical attributes and opportunities by means of the Assessment System, and a second design support phase that guides the selection of the best design strategies to eliminate, mitigate, or compensate for the factory's overall impact on the landscape, for which the Catalogue could provide inspiration. In brief, the project is assessed across the ten evaluation categories of the amended LEED scheme. Then, based on the result, the user defines the main objective and selects appropriate tactics to mitigate the negative impact, transforming them into specific design practices that are incorporated into the project.

The application process is iterative: because the topic encompasses many aspects that are difficult to quantify, the evaluator (whether an external or internal company professional) is invited to repeat the analysis-mitigation procedure more than once, implementing corrective measures at each subsequent application until a satisfactory level of compatibility with the landscape (i.e. score) is achieved. Figure 5.8 graphically illustrates this procedure, where the relationship between credits and tactics is made evident.[9]

In addition to companies and professionals, the toolkit implementation could be of interest to local administrations. For example, if they whised to promote territorial quality, they could offer incentives to enterprises who achieve a specific degree of compatibility. A public administrative body could also use the toolkit to select possible mitigation actions to be included in local planning codes; thus, it could be used as a checklist to judge a project and eventually provide suggestions and recommendations during the authorisation process.

Beyond these, other uses of the protocol, or a part of it, can be identified. In fact, its structure does not exclude it from only being accessed in some sections, or even reading/using a single tactic's card or credit.

To sum up, the toolkit is intended as a design support tool that promotes the achievement of high levels of sustainability in factory design assuming a holistic perspective. The instrument assists companies in the identification of the major impacts on the landscape at site, building and process levels; it then returns a score that expresses the overall level of compatibility of the project on the context and, in the end, encourages the implementation of best design practices that enable the highest grades to be achieved in the Assessment System.

[9] Specific links between tactics and credits can be found also within their cards. For instance, in the bottom part of the credit sheets, the name of possible tactics to be adopted is reported. In addition, each tactic reminds of the mother-credit by means of the alphanumerical code in its title, which is the same of the related credit.

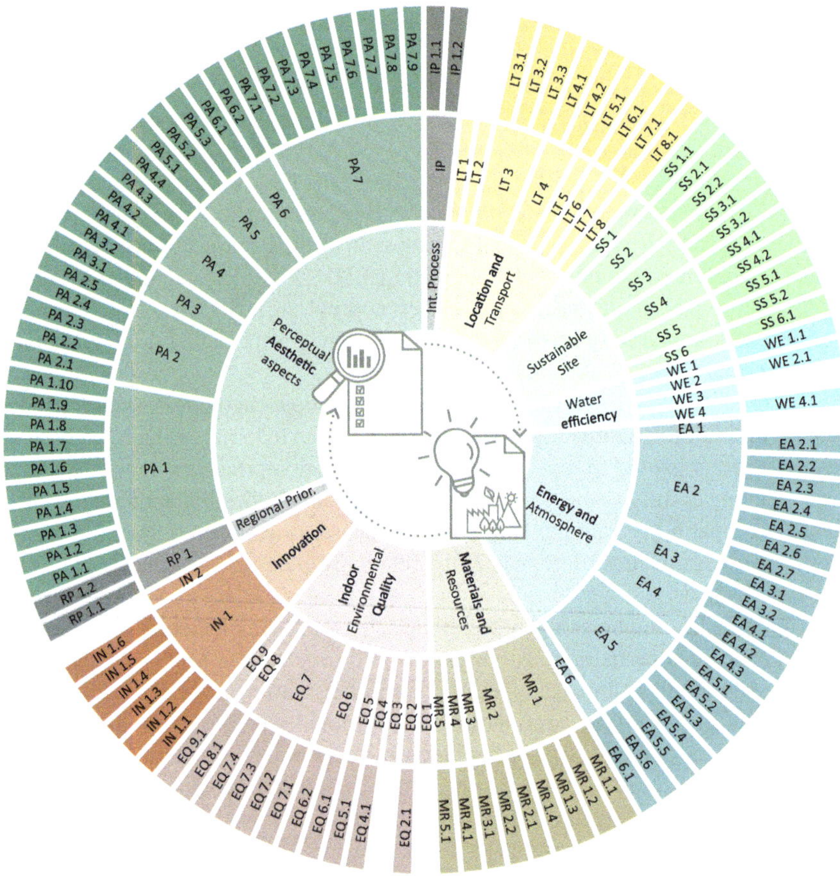


**Fig. 5.8** Reiteration of evaluation and design to increase compatibility and decrease residual effects. From the centre: credit category, credits, tactics

## References


Butler, A., & Berglund, U. (2014). Landscape character assessment as an approach to understanding public interests within the European landscape convention. *Landscape Research, 39*(3), 219–236. https://doi.org/10.1080/01426397.2012.716404.

Cassatella, C., & Peano, A. (Eds.). (2011). *Landscape indicators: Assessing and monitoring landscape quality*. Berlin: Springer. https://doi.org/10.1017/CBO9781107415324.004.

De Carlo, G. (2013). L'architettura della partecipazione (S. Marini (Ed.)). Quodlibet.

Doan, D. T., Ghaffarianhoseini, A., Naismith, N., Zhang, T., Ghaffarianhoseini, A., & Tookey, J. (2017). A critical comparison of green building rating systems. *Building and Environment, 123*, 243–260. https://doi.org/10.1016/j.buildenv.2017.07.007.

Marchi, L. (2020). Towards the harmonization of industrial facilities with the landscape. A catalogue of good practices. *SMC—Sustainable Mediterranean Construction, 12*, 156–160.

Marchi, L., & Antonini, E. (2021). A tool to evaluate what makes a factory sustainable. *International Journal of Sustainable Building Technology and Urban Development, 12*(2), 142–153. https://doi.org/10.22712/susb.20210012.

Marchi, L., Antonini, E., Orioli, V., & Evans, S. (2020). Mitigating the impact of factories on the landscape: An assessment and design support tool. *International Journal of Sustainable Development and Planning, 15*(3), 309–318. https://doi.org/10.18280/ijsdp.150307.

Rosley, M. S. F., Lamit, H., & Rahman, S. R. A. (2013). Perceiving the aesthetic value of the rural landscape through valid indicators. *Procedia—Social and Behavioral Sciences, 85*, 318–331. https://doi.org/10.1016/j.sbspro.2013.08.362.

Sowińska-Świerkosz, B. (2016). Index of Landscape Disharmony (ILDH) as a new tool combining the aesthetic and ecological approach to landscape assessment. *Ecological Indicators, 70*, 166–180. https://doi.org/10.1016/j.ecolind.2016.05.038.

Tolli, M., Recanatesi, F., Piccinno, M., & Leone, A. (2016). The assessment of aesthetic and perceptual aspects within environmental impact assessment of renewable energy projects in Italy. *Environmental Impact Assessment Review, 57*, 10–17. https://doi.org/10.1016/j.eiar.2015.10.005.

Tudor, C. (2014). An approach to landscape character assessment. In *Natural England* (Issue October). https://assets.publishing.service.gov.uk/government/uploads/system/uploads/attachment_data/file/691184/landscape-character-assessment.pdf.

U.S. GBC. (n.d.). *LEED pilot credit library*. Retrieved August 22, 2021, from https://www.usgbc.org/pilotcredits.

U.S. GBC. (2013). *LEED v4. Impact category and point allocation development process.*

U.S. GBC. (2014). *Reference guide for building design and construction.*

U.S. GBC. (2017). *LEED v4 for building design and construction.* http://www.usgbc.org/sites/default/files/LEEDv4BDC_04.05.16_current.pdf.

# Chapter 6
# Toolkit Implementation: An Italian Case Study

**Abstract** An implementation of the toolkit is presented here in order to provide a practical idea of its possible results, by means of an Italian case study (Orogel). The reader is guided through the application in the same way as the architect who was engaged by the company to propose mitigation scenarios for their brand-new facility in Emilia-Romagna. Thus, in this chapter it will be discussed: where to begin with the test; the kind of materials and information that is required; how to discover which impacts are most significant in the project; and how to select tactics and put them into action as specific design practices. For this case study, the iterative application of the toolkit led to three mitigation scenarios. From the first, which envisions a minimal retrofit operation achieving an overall good but not exceptional rating, to the third, which greatly improves the plant's overall compatibility with the landscape thanks also to a volumetric addition with outstanding performances. This chapter concludes with a discussion on general strategies that might be valid globally regardless of the specific case study.

**Keywords** Bioclimatic strategies · Design scenario · Visual mitigation · Warehouse · Worker's well-being

The development methodology and the main features of a toolkit for designing sustainable factories were described in the preceding chapter. Its general application procedure, namely the protocol, has also been described. This chapter will illustrate a sample application of the toolkit to assist the reader in gaining a clear and more practical understanding of the toolkit's usefulness and potential outcomes.

The Orogel case study is used to illustrate precisely where to start in the process, what basic information is needed, how to identify the most important project interferences that require mitigation or special attention, how to select relevant tactics and transform them into specific design elements, and how the iterative process should led to a progressively higher level of sustainability. So, the purpose of this chapter is more to guide the reader through the key steps of the protocol rather than to illustrate a specific case. Nevertheless, the three design concepts for reducing Orogel's impact on the landscape might also inspire the reader to adopt certain practices.



L. Marchi, *Designing Sustainable Factories*, Advances in Global Change Research 72,
https://doi.org/10.1007/978-3-030-92227-6_6

## Orogel Case Study

Orogel is an Italian agri-food company that has been leading the Italian frozen vegetables market for over fifty years. Over the decades it has also added fresh vegetables and jams to its array of products. As mentioned in previous chapters, the company has supported the Author's PhD research with a grant, and has also made itself available as a case study. This collaboration between Orogel and the University of Bologna demonstrates the innovative spirit of the firm, which is also expressed through it understanding of the importance of factory and headquarters architecture both as a marketing tool and as an example of responsibility for the protection and enhancement of the territory (Marchi, 2019).

As part of the agreement, Orogel provided the necessary resources to be used in this study, and asked for solutions to be developed for the integration of its industrial sites into the landscape while promoting the food culture. The wide range of functions, processes, and building structures that it has within the country make it an ideal case for demonstrating different research applications for future tests (e.g., new construction, complete renovation or retrofitting).

As a starting point, the complex chosen to illustrate the implementation of the toolkit is a brand-new facility that was under construction in Cesena (FC) at the time the study began (2016) and finished in 2019. For this reason, the version of the Assessment System for New Construction was used with a few exceptions. For example, tactics dedicated to NC that the company could no longer apply to the ongoing construction were not selectable.

Given that the study was not focused on earning a real LEED certificate, original LEED credits were assigned in an expeditious way,[1] whilst a detailed measurement of the PA credits was performed by the eight new info-sheets. On this basis, appropriate tactics to mitigate the impacts were selected and turned into design actions only at the conceptual level, not as an executive design project.

In order to obtain a finer level of compatibility between the site and the surrounding area, the process has been repeated three times, resulting in three possible mitigation scenarios. In this specific test, the three scenarios are not strictly sequential but rather projects which demonstrate a broad and diverse use of tactics. This means that each application may not include all the prior tactics and practices, but it may adapt and change strategies. This also highlights that there is no one predetermined best result (a universal panacea for bad factories or a beauty canon). The final scenario may depend on the design choice and preferences of the company or the practitioner who is implementing it, along with specific features of the context and project requirements.

The toolkit application is now discussed from the perspective of an architect (the Author) who has been engaged by the company.

**Background information**. Before starting the assessment, some general data on the firms is collected to frame the values, specific requirements, needs, and preferences of Orogel. In this case, the following is deemed relevant:

[1] This means that the score has been assigned to each credit by quickly consulting the User Guide, not precisely calculating those credits that require formulae and indexes.

- the company's *target market* is large, from mass retailers (thus serving end consumers) to other industries that use their products for secondary processes.
- The *economic dimension*, with its 289 million turnover in 2020 and 11 sites across the Country, the company is a leader in both the national and international market for frozen products.
- The company *values*. In the last decade the company has particularly invested in enhancing its image through a renewal of its brand and a marketing campaign called Orogel 360°. Environmental quality (including cropping systems, seasonality, and good nutrition) and social well-being are two of its main pillars.
- Concerning the environment, many awards attest to the great effort the company has been putting into reducing its impact on the ecosystem along the whole chain, from guidelines for associated farmers, to waste reduction in manufacturing processes, from energy efficiency within its facilities, to a new eco-friendly fleet of vehicles for transportation.
- Regarding the company's social initiatives, by means of a non-profit branch of its organisation—namely, *Fondazione Fruttadoro Orogel*—the company supports local socio-economic development. Social housing for workers and sponsoring a local football team are two good examples.
- Because Orogel products have a medium–high price tag, the company is pushed to add intangible value to its image so that customers are willing to pay a premium. This partly explains their interest in sustainability and in the consistency of the image of their facilities with their values.

**Introduction to the site**. Orogel had to extend its historical production complex in Pievesestina (Cesena, Italy) (Fig. 6.1) as they needed to restructure and make their factories more efficient, but also because of their extraordinary market growth that required them to create more storage spaces.

Besides Orogel 1 and 2, the company opted to construct Orogel 3, a large new site including two buildings (plus one in the future) for packaging, one (plus one in the future) big automatic cool warehouse, and a jam manufacturing plant.[2] As a result, the firm's storage capacity is increased, and geographic fragmentation reduced, limiting additional costs associated with the rents of third-party warehouses scattered thorough the country. Meanwhile, by moving some functions from Orogel 1 and 2 to the new Orogel 3, there will be more room, in the medium-long term, to renovate the existing assets in terms of functionality, visual quality, and technology.

The first phase of this complex project (Orogel 3) was completed in 2019, as seen by the promotional video (Orogel 3, n.d.), but the toolkit has been applied to the entire site.

**Assessment stage: baseline scenario (Fig.** 6.2). A territorial, morphological, and constructive analysis of the complex is performed to measure its impact. The credits that most help to evaluate the site quality are SS1 and PA0, both Site assessment: the first from the environmental point of view, the second from the perceptual-aesthetic

[2] The project approved by the Municipality of Cesena can be searched online as "Piano Urbanistico Attuativo di Iniziativa Privata Area Pregresso PRG 85 n. 25 Pievesestina Via Dismano e relativi progetti esecutivi delle opere di urbanizzazione e degli edifici 1, 2, 3, 4 e 5, approvato con delibera di Giunta Comunale n. 91 del 05/05/2015.".

**Fig. 6.1** Bird's eye view of Orogel in Pievesestina, Cesena (Italy), 2017 [Google Maps]

one. , A series of issues emerges from this analysis, the most important of which are discussed here.

From the territorial point of view, important contrasts occur. The company must cede a part of the area to the municipality for the creation of urban infrastructure such as cycle lanes and public car parks along Via Dismano (longitudinal axis of the area). As a result, new buildings were positioned back from that area, more adjacent to the E45, an important national highway which is also elevated along that section, resulting in the factory being highly visible from this second road. This fact brings about an unwanted effect of "showcasing" the new site from the highway and, given that it is not "visually pleasant", it exacerbates a negative public perception of the site in the landscape, contrary to those values held by the company. At least in part, a different location of the production site could have reduced its visibility. However, even with this limitation, another configuration of the complex could have improved this problem, while still respecting its functional and logistical requirements.

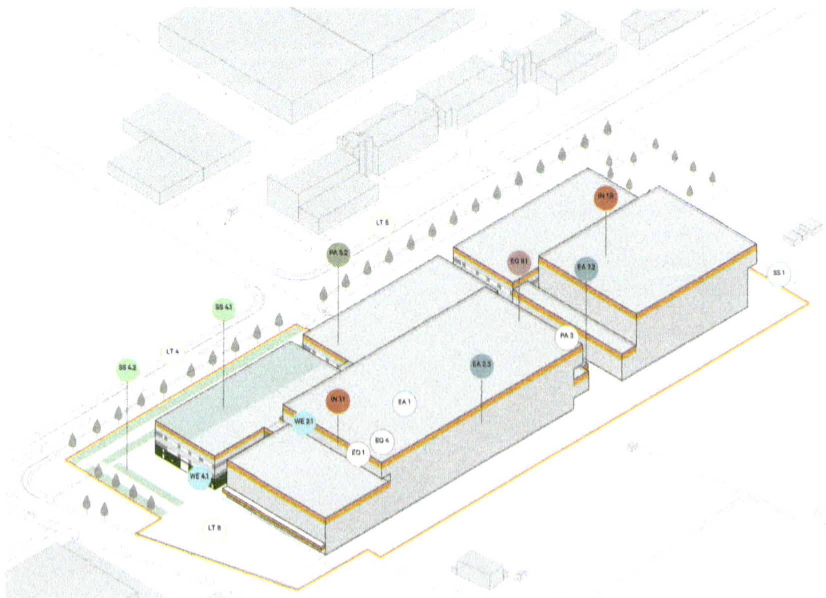


**Fig. 6.2** Baseline scenario. Already implemented tactics are represented by coloured balls

Concerning the architectural consistency of the complex, the massive size[3] (the first warehouse is 120 m long by 30 m high) compared to nearby constructions, as well as the anonymous, plain treatment of its facades, and the lack of social attractiveness for locals, emerge as the main critical issues. As for the chromatic features, a bright yellow strip encloses the white aluminium cladding at the top: on the one hand this evokes the colour in the firm's logo, which could be seen as positive, but on the other, it clashes with the structure's natural surroundings.

In addition, the quality of open space within the complex is poor: space is mainly devoted to service areas for trucks, hardscapes cover the entire site, with only limited parking on permeable surfaces; the limited green area is mostly composed of residual spaces, and traditional tree lines are planted to shade the car parks.

In terms of technology, the new buildings appear to be highly innovative and cutting-edge as several strategies have already been implemented: from envelope energy efficiency[4] to water conservation, from the reuse of process water[5] to the

---

[3] Freezing of seasonal vegetable products entails storing the product at the time of harvest in sufficient amounts to cover the entire year supply. As a result, these structures have exceptionally large dimensions, which might be difficult to mitigate if not properly designed.

[4] For example, ventilated façades and highly insulated panels.

[5] The installation of a biological purification plant system allows portion of the water used in the manufacturing process to be reintroduced after being micro-filtered and utilized for machine washing (the legislation does not allow it to be used for cleaning food).

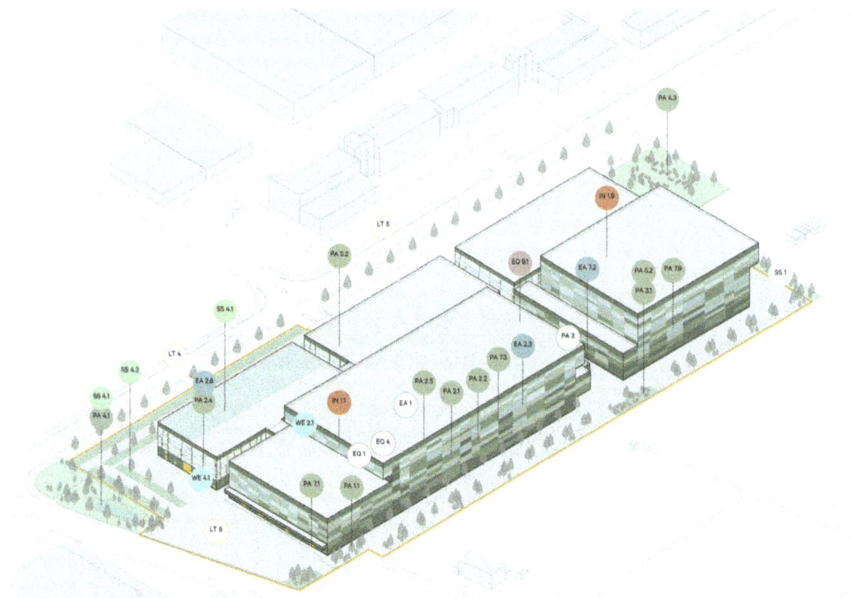


Fig. 6.3 Scenario one

reduction of atmospheric emissions. Despite this, their energy and resource demands are still significant.[6]

With these premises, the firm faced the difficult issue of successfully and directly communicating the implemented best practices to clients, staff, and locals of the complexes closest to the settlement. Local populations, on the other hand, have expressed their dissatisfaction with the degradation of their landscape by these massive, impersonal constructions, raising concerns of visual and perceptual integration.

The overall score of the project was 38 out of 127 points, performing better in the energy and atmosphere category, and worse in the perceptual-aesthetic ones.

**Scenario one: cosmesis (Fig. 6.3).** Given that the overall environmental performance of the buildings within the complex was good, the first hypothesis focused on improving the integration of the complex into the scenery while enhancing just a few environmental measures. PA 2.2 (Pattern from context) and PA 2.5 (Graphic design), for example, are used in green-palette areas of the envelope evoking both nearby fields and natural vegetation. Darker green is utilised towards ground level to focus the visual mass of the building, becoming lighter and transitioning to a light blue

[6] Refrigerated warehouses must maintain an ideal operating temperature of around −25 °C inside, so the energy demand remains significant even with the most stringent efficiency standards and that, to be met, necessitates the construction of large power plants that have an impact on the landscape. This still happen despite opting to maintain the product at −22/23 °C to save electricity and money, as Orogel did.

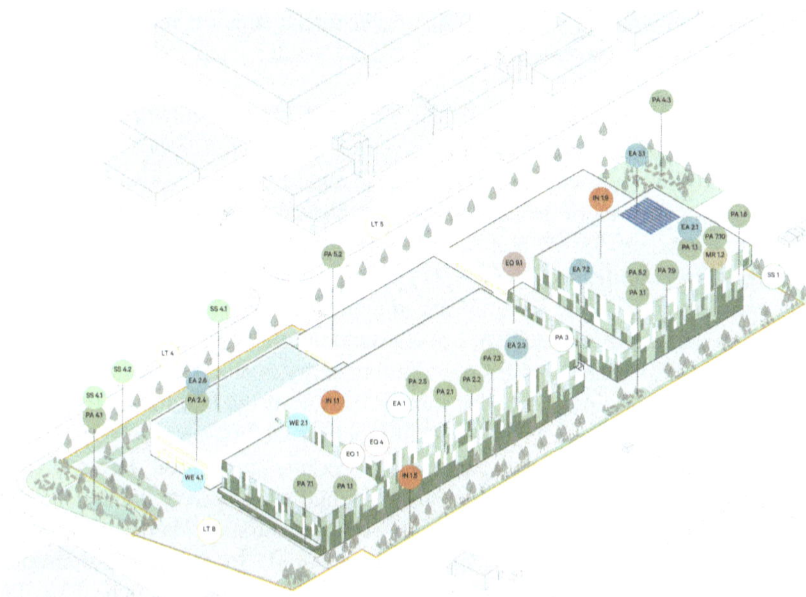


**Fig. 6.4** Scenario two

sky towards the top. A consistent addition of trees PA 1.1 (Visual screen) and green space PA 4.3 (Public Park) is also suggested. Then, PA 4.1 (Multifunction facility) is combined with SS 4.1 (Rainwater collection) resulting in a specific design practice that makes use of a ground depression, serving as a lamination basin during heavy rains, and as a park for workers and neighbours the rest of the time.

The first scenario scores 42 out of 127 points, thus it can be compared with the LEED level "Certified". This suggests that the improvement is acceptable but not outstanding.

**Scenario two: shielding and second skin (Fig.** 6.4). The first scenario demonstrated that working on the perceptive and visual quality of the building is good, but the overall benefit is still limited. The intermediate scenario attempts to elevate the overall level of integration of the building with the landscape by combining environmental, social, and perceptual practices.

The main design feature is a set of curtain walls and visual screens that both mitigate the visibility of the complex and accomplish environmental functions affecting the indoor and outdoor micro-climate (e.g., shading, ventilation cavity, solar energy production): EA 2.6—Green Façade is integrated with EA 2.3—Thermal insulation, and more "aesthetic" ones as in PA 1.1—Visual screen (both light blue and green coloured panels and tree lines). In turn, the screen layout helps to make the building more sympathetic with its surroundings by picking up the colours and patterns of nearby fields (PA 2.1—Colour from the context, and PA 2.5—Graphic design). The enhancement of the environmental performance is also helped by EA 5.4—Cogeneration; while effective water management is reached by means of WE 4.1—Water management systems, SS 4.1—Rain water collection and SS 4.2—Pervious surfaces.

Social-based tactics are also utilised, such as PA 4.1—Multifunction facility and PA 4.3—Public Park that boost the attractiveness of the site for workers, visitors, and local communities. Specific practices in this regard might be a small garden for workers, a park with water effects, and outdoor sports equipment.

This scenario scores 57/127 points, comparable with a LEED 'Silver' certificate.

**Scenario three: additional volume (Fig. 6.5).** For the last scenario, sustainable holistic practices on existing facilities are implemented. For example, a green roof that improves the micro-climate (EA 2.5—Ground thermal properties and EA 2.4—Thermal mass) is combined with EA 5.1—Solar energy as photovoltaics on the large roofing and SS 4.1—Rainwater collection, on the top of the building. Green space design SS 3.1 is mixed with PA 4.1—Multifunction facility, arranging spaces for cultural events, pocket parks and gardens for workers' recreation during breaks, and PA 4.3—Public Park. Many other environmental actions were implemented to mitigate the negative effects emerged in the Assessment stage, but still important non-mitigable environmental and social issues occured. So, they are proposed to be compensated for by a volumetric extension that hosts the company headquarter, while visually mitigating the complex by shape and colours. In fact, the green sinuous roof evokes close hills and farming, as does the vertical treatment of the façade of the warehouse (PA 2.3—Vegetated roof and PA 1.6—Reference to the skyline)

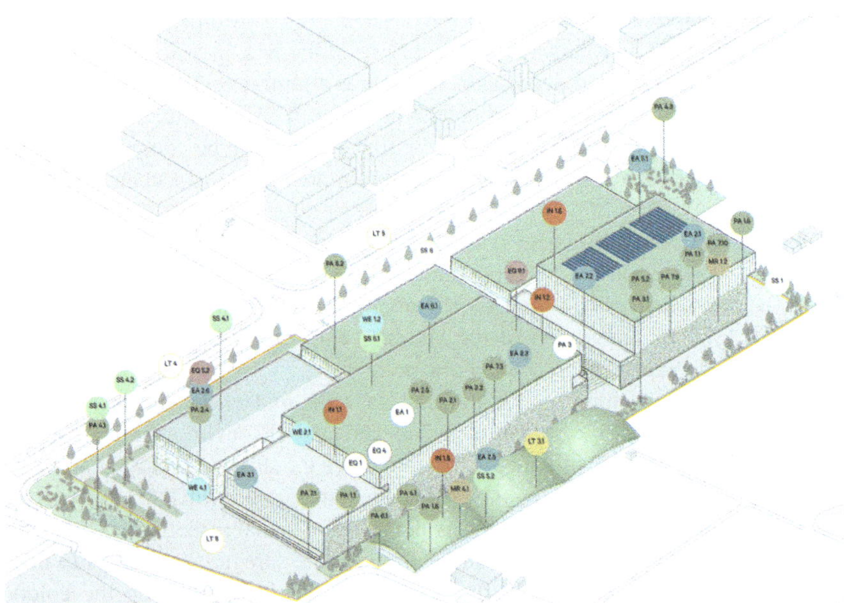

**Fig. 6.5** Scenario three

and the light green façade devised for the smaller buildings along via Dismano (by means of EA 2.4—Thermal mass, EA 2.1—Ventilated façade and EQ 5.1—Natural ventilation). This latter implementation consists of white semi-opaque cladding; this recalls ice, evoking the frozen vegetables process, while filtering the daylight inside.

The use of green, vegetated materials are a reminder of Orogel's raw materials (agri-food products), according to the tactic PA 7.2—Reference to product's ingredients.

All in all, the scenario results in 79 out of 127 points, comparable to LEED 'Gold', the second-highest level.

**Reflection on the case study outcome**. The application of the toolkit provided the firm with an assessment of its level of impact on the landscape, along with detecting the main issues that need to be solved. In addition, the company learned about practical actions that can be implemented to mitigate certain problems and improve the overall compatibility of Orogel 3 with the context while improving their corporate image.

The tactics cards that were provided to the firm describe both tangible and intangible benefits associated with certain actions, as well as many real-world examples and references. Therefore, the company needs "only" be motivated and put these ideas into action.

Unfortunately, the specific outcomes were influenced by former decisions that prevented the scenarios from receiving higher scores. For instance, an alternative shape of the building that was still consistent with production requirements could have had a better visual impact. Similarly, alternative placement and layout of the buildings on the site could have had a positive effect on their surroundings.

These are just a few personal opinions that support the idea expressed in the previous chapters that sustainability cannot be considered as an add-on. Only by incorporating systemic thinking and integrative processes early in the design stage can a project be successful at little or no additional cost. Otherwise, post-construction mitigation activities will inevitably represent an extra feature, and therefore an extra cost, even if these are welcome when no other possibilities exist, such as in the case of retrofitting/renovating the existing asset.

## From the Case Study to General Strategies

In Chap. 4, some insights into patterns of good practices were presented and it was argued that even though they have been found in agri-food companies, they are most likely also relevant for other industrial sectors, as well as residential, commercial, and various other functions. The same is true for tactics from the catalogue, including those that have been implemented in the Orogel case study, which can be easily transferred to other fields. Thus, many combinations of tactics are suitable for "universal" application, regardless of the specific features, location, and function. Then, it is the specific practice that must be adapted to the case.

Even more universal are some general strategies to which some tactics contribute. Although they belong to topic-specific categories, there are mutual relationships between certain tactics, both from within the same category and from different categories, that make it easier to take a comprehensive approach to sustainable factory design. In other words, an "environmental" tactic will likely have social and perceptual-aesthetic consequences that will strengthen a certain design strategy. This is a point which the whole book stresses.

Some examples of these synergies may be helpful to the reader, starting from the multifaceted classification of tactics (ref. Chap. 5).

*Bioclimatic strategies*—that is a possible attribute of the facet "Strategy" within the tactics cards—can be applied in a variety of ways and through different methods. Solar energy, natural ventilation, daylight, or water evaporative processes are just some potential practices. Regardless of the specific choice, the result is that the energy demand of the building is lowered due to the use of these natural means. First of all, this impacts the environmental balance of the building; but also the health and well-being of its occupants, whose indoor and outdoor comfort are improved through natural techniques. Not to mention the building's lower operating costs due to reduced energy consumption. Several tactics refer to these general strategies, involving multiple scales and issues: SS 1.1—Site ventilation and cooling; SS 3.1—Green space design; SS 5.2—Green folded surfaces; EA 2.5—Ground thermal proprieties; EA 2.6—Green façade; EQ 4.1—Natural ventilation; PA 1.3—Landform; PA 2.3—Vegetated roof.

*Visual mitigation* works similarly, even though, at first glance, it might appear to be merely a perceptual-aesthetic issue. Instead, numerous actions can be taken to reduce the factory's visual pollution in the context and at the same have a beneficial impact on the environment, the workers, and the nearby neighbourhoods. As mentioned in this book, visual mitigation does not always imply that the structure is hidden or made mimetic, but it can also mean that the perception of the building is enhanced or even highlighted using colour, graphic, materials, shape, and other architectural techniques. Contributing tactics are for example: SS 5.1—Green roof; SS 5.2—Green folded surfaces; EA 2.1—Ventilated façade; EA 2.6—Green façade; PA 1.1—Visual screen; PA 1.2—Horizontal dimension; PA 1.3—Landform; PA 2.1—Colour from the context; PA 3.2—Technological barrier. At the same time, benefits within the social sphere can occur, for example, by combining visual and noise mitigation (PA 3.2), improving the light and the visual comfort of the building's occupants, and by offering workers visual distractions with green, natural, and relaxing colours and materials (PA 5.2). Accordingly, positive effects within the environmental sphere are easy to estimate, such as improved thermal mass that reduces energy demands for cooling in the summer (EA 2.4) or a better outdoor micro-climate thanks to evaporative cooling (SS 5.2).

*Worker's health and well-being* is another general strategy that can simultaneously contribute to multiple aspects. For example, not only can PA 5.1—Services for workers and PA 5.2—Ergonomic workplace benefit employees, but also the company itself by increasing workers' loyalty and productivity (U.S. GBC, 2014). Similarly does PA 4.4—Visibility and identity of open space, which aims at offering workers pleasant spaces for relaxation but can also contribute to enhancing the overall quality of the site, and hence the factory's image in the eyes of locals, visitors, and even car drivers.

To conclude, the idea that lies behind the book and toolkit alike—i.e. which synergies among different landscape dimensions are beneficial to the successful design of sustainable factories—is well synthesised in these multiple links and interactions among tactics. Mentioning the Italian writer G. Rodari in his *The Grammar of Fantasy: An Introduction to the Art of Inventing Stories*:

> A stone thrown in a pond gives rise to concentric waves that spread over its surface, involving in their motion, at different distances, with different effects, the water lily and the rod, the paper boat and the fisherman's float. Objects that each stood on their own, in their own peace or in their sleep, are as re-called into life, forced to react, to enter into relationship with each other (Rodari, 1973)[7].

As happens in a well-written tale, the same is true in this field: various parts of the context, the firm, and the manufacturing process must interact and cooperate to provide a successful, sustainable, and attractive project. A successfully designed sustainable factory is no different; it will generate several ripple effects into its surroundings inspiring others to follow the same path, or at least to reflect, to be curious on the kind of benefits that the company has gained.

## References

Marchi, L. (2019). Orogel. Tra radicamento territoriale, innovazione tecnologica e immagine aziendale. In F. Gulinello & E. Mucelli (Eds.), Fabbrica della meraviglia (pp. 9–14). LetteraVentidue.

Orogel 3. (n.d.). Retrieved August 22, 2021, from https://www.orogel.it/news/orogel-news/orogel_3_il_nuovo_complesso_tecnologico.

Rodari, G. (1973). *La grammatica della fantasia.* Einaudi.

U.S. GBC. (2014). *LEED core concepts guide.* U.S. Green Building Council.

[7] The original Italian text is «Un sasso gettato in uno stagno suscita onde concentriche che si allargano sulla sua superficie, coinvolgendo nel loro moto, a distanze diverse, con diversi effetti, la ninfea e la canna, la barchetta di carta e il galleggiante del pescatore. Oggetti che se ne stavano ciascuno per conto proprio, nella sua pace o nel suo sonno, sono come richiamati in vita, obbligati a reagire, a entrare in rapporto tra loro».

# Chapter 7
# Conclusion

**Abstract** The book has attempted at demonstrating that the opening image of factories harmonised with the landscape is not utopia. Rather, if a holistic approach to sustainability is applied from the early design stage, both in new buildings and in retrofit of existing facilities, good practices that help to reach that scenario can be implemented at little or no additional cost. Assuming the broad concept of landscape promoted by the European Landscape Convention, the study addressed the issue by considering environmental, perceptual, and social aspects as complementary facets of the place. On this basis, barriers and drivers of the change have been discussed, as well as tools that might support companies in designing more sustainable factories. Then, a toolkit for the assessment and mitigation of factories' detrimental effects on the landscape has been illustrated, along with its test on an agri-food case study. After discussing the toolkit limitations and further developments, the chapter opens to future scenarios before concluding that now it is time to act.

**Keyword** Holistic sustainability · Win-win logic · Industrial landscape · LEED

The intent of the book was to discuss why and how to design more sustainable factories. The reader has been taken on a journey beginning with the reflection that industrial sites more respectful of the landscape are needed. Then, both well-known and unexpected gains of this essential transition were discussed, and it was argued that the change would benefit many of the stakeholders (e.g., firms, policymakers, local communities). In the end, the reader was illustrated how companies can be supported in this pathway, especially by means of a toolkit to mitigate detrimental effects of their industrial sites, thus respecting the environment, the scenery, and the local community, while promoting their competitiveness.

In the first chapter the basic assumptions of the study are illustrated, among which the concept that all the territory is landscape as defined by the European Landscape Convention (ELC) and that all its facets are important to define the quality of a place and the identity of its community. On this premiss, the chapter examines the multi-scalar and multi-faceted negative effects that industrial sites have on the context, which are primarily influenced by the location, site quality and technological features, construction techniques, aesthetic attributes, and resource management in general. It emerges that more sustainable factories are needed not simply to meet global

 
L. Marchi, *Designing Sustainable Factories*, Advances in Global Change Research 72,
https://doi.org/10.1007/978-3-030-92227-6_7

targets for energy and resource sustainability, but also to ensure workers' well-being and the quality of people's everyday landscape. The outline of the emerging policy framework that supports this change at both global and European level concludes the introduction.

The second chapter provides an overview of how the impact of factories on the landscape has changed over time becoming progressively heavier. At the same time, the factors that have influenced the relationship between the site and its surrounding are discussed. Throughout history, economic issues have been a major impact on factory design. The increased awareness about the global environmental crisis has prompted a shift in direction, but the understanding of the value of sustainable factories remains low, and the importance assigned to costs still outweighs perceived benefits. Gradually also the concept of visual pollution has taken place and in some cases it helped to improve the image of the factory, especially when linked to brand identity projects. However, while big enterprises find it simpler to invest in sustainable factory design and in green marketing, small and medium businesses struggle with resource shortage and lack of knowledge. As a result, the great majority of industrial sites that constitutes our daily landscape keep contrasting without any mitigation.

The factors that may encourage businesses to build environmentally and socially sustainable manufacturing facilities are discussed in Chap. 3. In addition to direct economic benefits that many may already be aware of (e.g., reduced operational costs connected to energy efficiency), other less common drivers are discussed. Among these, the increased competitiveness that companies can gain through a better factory image working synergistically with good environmental and social practices. In order to support awareness raising and concrete actions being undertaken, some tools have been developed to assess the impact or guide the adoption of good practices. However, the chapter concludes that they are sectorial and not effective in supporting the design of holistic sustainable factories, which instead has been demonstrated to provide interesting wins.

In Chap. 4, the example of agri-food as a sector that recognized the potential of sustainable factories earlier than others is given. The case indicates that the core assumptions of the book are not just theoretical, but also proven in the real world: otherwise, no company would have implemented the outlined best practices. Six exemplary patterns with references are shown at the end of the chapter. Although the focus is on agri-food because of its distinctive traits, several insights might apply to other sectors alike.

As contribution of the study to the research field, a toolkit to assess and design more sustainable factories is presented in Chap. 5 with the goal of assisting firms—supported by practitioners—in better recognising their impact on the landscape and detecting appropriate mitigation actions. Based on an iterative application procedure, the tool is made up of an Assessment System and a Catalogue of good practices. The first is a revision of the U.S. GBC LEED scheme, which evaluates the factory in several categories ranging from site quality to water management, from energy and atmosphere to materials, from indoor environmental quality to perceptual-aesthetic aspects. The second is a collection of virtuous case studies from which design/mitigation practices have been retrieved and, in turn, a library of general tactics has been devised. The latter represents the hinge of the whole mechanism, as it links practices from the Catalogue with credits from the Assessment System. Links and interconnections between the elements (i.e., tactics; practices; credits) are emphasized as being even more important than the single design choice.

Then, in the sixth chapter, a case study (Orogel 3) is taken as chance to illustrate the toolkit application. The reader is put him/herself in the architect's shoes. Three scenarios are presented, each with a progressively higher level of sustainability, followed by a discussion of the specific application. Beyond peculiarities of the case, the section ends with observations on the toolkit potentialities and some broader strategies that can be universally valid.

To conclude the book, a few more words on the toolkit's limitations and future developments follow.

The book illustrated and discussed the result of a research project aimed at providing companies with a design-support tool to eliminate or mitigate the impact of their factories on the landscape. The study has proven that a multidisciplinary approach is necessary to ensure firms economic benefits besides environmental and social wins for the community. So, in order to fill a gap in tools for assessing and mitigating the impact of industrial facilities on the landscape, the choice was to start from instruments that were already conceived for combining multiple factors and disciplines around sustainability, while being user-friendly and market-driven. Thus Green Building Rating Systems (GBRSs) were chosen as reference, since they can integrate the evaluation of diverse criteria into a consistent and easy to understand result.

Because GBRSs are primarily targeted on environmental impact, the idea was to add a new category addressing perceptual-aesthetic aspects, being these latter largely neglected in factory design but increasingly important to both the company and the community. This choice was not easy going, as combining respectively quantitative and qualitative indicators into a coherent framework is a challenging task, which is partly the reason why researchers have addressed the problem per sectors up to now. Nevertheless, it appears to be the only way possible to properly approach the issue. As a consequence, a certain degree of flexibility is expected in the expert's judgment of the 8 PA credits. Hence, a toolkit's pilot phase (yet to be defined) might be helpful to validate the instrument. Should the results of this test proven to be too variable and subjective, more stringent gauges could be implemented in a future version.

Another element that might seem risky—at a first glance—concerns with the idea of designing more sustainable factories starting by single elements—such as good practices—to be combined according to the user preferences and abilities. One of the key hypothesis of the work was that a holistic approach to landscape compatibility of factories is needed to both address the topic in its complexity and take advantage of mutual relationships. Instead, a tool that breaks down the problem in several parts and topics is provided to the user. As much as it may appear controversial, I think that the approach well recalls the idea of "assembling an image" by Yona Friedman (Friedman, 2008): it comes of putting together single elements (*assembling*) to create something that is unitary as we perceive it (*the image*). In fact, while dogs have a holistic interpretation of the world (vision, in physical terms), as humans we have developed an analytic interpretation of the world and we are capable of split it up in its basic components, each of which matter for the final output. So, it is important to underline that the toolkit helps to analyse the reality—breaking it down—and thus guides the user in identifying main criticalities and offering her/him an inspiration. Thereafter, it is their job to assembly the parts in a consistent way and deliver a successful project. This is to say that the fact of applying the toolkit does not ensure by itself to obtain a good project. The same "score" can be achieved in several different ways. As Friedman suggests, for instance, 5 can be obtained by $0 + 5$, $1 + 4$, or also $4 + 1$, $5 + 0$, $6 - 1$, and so on. Likewise several approaches and design choices can lead to a sustainable factory, but then the relationships and balance among the specific measures—proper of each case, its requirements, and features—make a good design.

More in general, concerning future enhancements and developments of the toolkit, several opportunities have arisen across the study.

In terms of content, the instrument is potentially extendable in size and scope. In fact, the case studies collection can be enlarged at any time, because the structure is adaptable. This in turn might broaden the list of practices, and the library of tactics alike. Similarly, other industrial sectors could be considered to furtherly widening the applicability of the tool.

Furthermore, a concrete exploitation of the Assessment System by a Green Building Council could be envisaged. As illustrated in Chap. 5, despite being a versatile tool for adaptation, each change to a LEED scheme must undergo a member consensus-based but rigorous approval process. For this reason, an easier stage has been envisaged in the study by treating some of the developed PA credits as Pilot Credits. This could be also the simplest approach to include them in LEED for real. However, in order to fully exploit the tool's potentialities, the addition of the entire new PA category in the actual LEED is recommended. This certainly necessitate an established interaction with the GBC and more specialists working on the tool, which can currently serve as a solid base for getting started but is not yet ready to market. Thus, if for instance the GBC Italia would be interested, the actual implementation of the PA category in LEED could be explored, as well as the development of one or more Pilot Credits.

Regarding the tactics, in technical terms, the library has already been predisposed to become a digital atlas, likewise some well-known architectural web portals where case studies can be looked for by multi-faceted search (e.g., ArchDaily, Divisare, The Plan). Even if it currently lacks the impact assessment section, the digital version of the tool would make it easier to navigate and find mutual relationships and synergies among the tactics. Up to date, the Author has defined the site's architecture but has yet to find a digital developer.

Beyond presenting the toolkit, some general reflections and prospects have arisen in the book. They all refer to a "sustainable factory" as a construction that has a low environmental impact, is well integrated in the surrounding, and provides a healthy and stimulating working and living place.

**Holistic sustainability**. The idea that in some cases 1 + 1 do not make 2, but more, is central. That is to say that adopting two design practices that work simultaneously can reduce the required effort to achieve the same performance; or conversely that the result can be a whole better than what could have been obtained through the algebraic sum of two elements. The reference is to those positive interactions that can make designing or retrofitting a factory in a sustainable way beneficial to the company, by boosting the potentiality of two tactics: actions that complement each other creating a whole greater than the sum of its parts. A green hill next to the building with a public park on the surface, for example, helps to increase the building's thermal inertia, to improve the site's micro-climate and visual mitigation, as well as to enhance its attractiveness. In contrast it would not have been as much appealing having just a green roof on the top of the plant and a park at ground zero. This in turn can result in reduced operational costs and even improved competitiveness of the company through a more pleasant architecture.

The argument supports the concept that sustainability can be a core value for the business, rather than an extra feature. To this purpose, it is recommended implementing systemic thinking and integrative processes—as described in Chap. 3—to find synergies during the design stage, whether or not the toolkit is used. In fact, these methods allow the designer or manufacturer to focus on the connections instead of the single pieces, hence to find the best compromise according to their needs and preferences. The concept is in line with other mainstream approaches around the built environment, such as resilience and life cycle thinking, and it is critical that all the people involved in the retrofitting or new factory design process are aware of them, so to facilitate a change of perspective.

**Win–win situation**. Not only is the idea of holistic sustainability linked with companies' interests, but also with a strong collective responsibility. The book has discussed how businesses, locals, and the territory more in general have all a vested interest in hosting a sustainable factory. Tangible and intangible benefits for firms should now be quite clear to the reader, as well as the value that an attractive, environmentally friendly, innovative company gives to the territory. Positive impact on workers, building's occupants and those who live nearby should be clear alike, mainly in terms of health and psychophysical well-being, but also in terms of social acceptance and appeal of the place. Intrestingly, the phenomenon of inverse-NIMBY (discussed in Chap. 3) has the potential to encourage the design of more sustainable industrial plants: that is, the idea that a good factory (environmentally and socially sustainable) can be even desired from locals who become proud of having such a landmark of territorial quality, attracting locals, visitors, and even consumers. So everyone is willing to have "that" extraordinary winery close to home and claims: Please, In My Back Yard! In this framework, positive impact factory or regenerative projects that go beyond the mere negative impact mitigation by adding new value or services to the area, represent a stimulating field for future research and experiments. Finally, this positive circle would represent a win–win situation for all. It wins the public if the private win, and vice versa.

**Need to rehabilitate the value and the quality of ordinary landscapes**. For decades architects have addressed the industrial sector asset only for what regards "industrial heritage". Therefore, great, famous interventions have been undertaken to restore derelict skeleton of industries from XIX or XX centuries, which were brought back into new life as parks, museums, commercial centres (i.e., adaptive reuse of factories). The same noble was regarded designing companies' headquarters because of their extraordinary communicative potential. Instead, in very few cases practitioners have addressed the design of high-quality factories.

Hence, while architects were dealing with other issues, the landscape has been populated by thousands of small and medium size industrial facilities detracting its quality. Even though it seems to be less appealing, it is time to act on this "ordinary" asset as well. Over two decades ago, Member States signed the ELC, sharing a new and more comprehensive way of thinking about and intervening in the landscape. Industrial complexes belong to those everyday landscapes which are recognised as deserving the same consideration of high-quality scenes, because they are a part of most people's everyday setting, and so affect their quality of life and essence as humans. This may appear to be too enthusiastic, but I truly think that the form and the quality of our environment contribute to shape our identity. As architects and businesses, we all have a challenging task ahead of us.

> "It seems to me," responded don Quixote, "that you aren't well-versed in adventures—they are giants; and if you're afraid, get away from here and start praying while I go into fierce and unequal battle with them." And saying this, he spurred his horse Rocinante without heeding what his squire Sancho was shouting to him, that he was attacking windmills and not giants. But he was so certain they were giants that he paid no attention to his squire Sancho's shouts, nor did he see what they were, even though he was very close (de Cervantes, 1992).

In the end, I am sure that many of us—like don Quixote—have always thought of factories as nasty giants. This book has sought to demonstrate that they could be "nothing more" than good and pleasant windmills.

## References

de Cervantes, M. (1992). Don Quixote (P. A. Motteaux (Translation) (ed.)). Wordsworth Editions.

Friedman, Y. (2008). L'ordine complicato. Come costruire un'immagine (L'ordre compliqué et autres fragments), Trad. it. Tramannoni 2011. Quodlibet.

# Afterword

The mission of the Green Building Council Italia (GBC Italia) is—similarly to many other GBCs around the world—to promote the diffusion of sustainable buildings' culture; to raise public awareness on the impact that the quality of the building sector has on the citizens' life; to provide clear indicators and methods to measure sustainability to the stakeholders of constructions; and to support the dialogue between workers in the building sector, by creating a community of sus-tainable constructions. We strongly believe that is possible to allow everyone to live in an environment that respects the needs of man and the ecosystem.

One of the mean we have to support this mission is to promote Green Building Rating Systems as tool to measure buildings' sustainability by multi-criteria assessment, and encourage the adoption of environmentally, socially and economically sustainable practices in design, construction and operation of buildings. In other words: by providing tools that drive the sector to be greener. Hence, GBC Italia encourages the change of the Italian market through a third-party certification system and its own certification protocols (GBC systems) that are specifically developed for the needs of the Country.

In this framework, several adaptations of international schemes such as U.S. GBC LEED have been developed in the last decade: GBC Home, GBC Quartieri (i.e. neighbourhood), GBC Condomini (i.e. multi-propriety buildings), and GBC Historic Building. The last, for example, is conceived to address the conservation, redevelopment, or renovation of national cultural heritage with a sustainable perspective. The idea arose from the desire to combine and merge LEED sustainability criteria with the excellent Italian knowledge about preservation of historical buildings. In practical terms, GBC Historic Building was developed by adding an entire new Credit Category—namely, "Historical Value"—which was reckoned to be necessary to comprehensively approach the refurbishment of Italian historical heritage and adapting to this end the classical LEED thematic areas and the relat-ed prerequisites and credits system.



L. Marchi, *Designing Sustainable Factories*, Advances in Global Change Research 72,
https://doi.org/10.1007/978-3-030-92227-6

On the same wavelength is the research project that underpins this book, whose early developments I heard about in 2018, when Professor Ernesto Antonini (supervisor of Lia Marchi) kindly invited me to participate in one of the periodic presentations given by PhD candidates at the Department of Architecture, about the progress of their studies. One of the starting points of the research was that the presence of badly designed and poorly performing factories has a significant impact on the Italian landscape, which is internationally renowned for its quality. Thousands of small to medium size plants, regardless of their surroundings in terms of natural, sociocultural, and perceptual-aesthetic values, are scattered throughout the country as the result of mere profit-making logics. Throughout history, factory design has simply responded to functionality as a fundamental requirement. As a result, nowadays there is an enormous, diversified asset that requires extensive rehabilitation, refurbishment, or retrofitting. Improvements in facility environmental and social performance must increasingly be linked to long-term economic sustainability. Indeed, the long lasting economic crisis that has made it difficult for businesses to integrate sustainability into their operations has been exacerbated by the pandemic, which has increased the urgency of businesses' economic sustainability while also highlighting the need for a greener working and living environment.

In this context, the book highlighted that neither companies nor architects/designers have tools that can assist them to address the issue, thus to effectively designing more sustainable factories. Therefore, assuming the complexity of the topic, the many and diverse aspects involved, the study developed a toolkit to assess the impact of factories on the landscape, and propose a catalogue of impact mitigation strategies.

The Assessment System, in particular, was built as an amendment of LEED, GBRS chosen for its potentialities, ease of use, diffusion and facility to be modified. In addition, the proposal of a LEED for factories is not totally new: since the 2012 the U.S. GBC has a LEED User Group: Industrial Facilities (LUGIF) that attempts to share opportunities and best practices for sustainable factories; GBC India has launched the first LEED scheme for factories; and not of minor importance hundreds of industrial plants have been certified via general LEED schemes at present days. Nevertheless, the book has pointed out that a sustainable factory is not only focused on "energy-environmental" approach, but it needs the other dimensions to be addressed. In this sense, a suitable and extensive application of LEED to factories is still lacking and the proposal tries to fill the gap by adding the new category.

The idea of a new Credit Category focused on Perceptual and Aesthetic aspects of the landscape is not so far from what GBC Italia has made with GBC Historic Building, by introducing the category "Historical Value". However, from the operative point of view, the addition of a new category requires a huge scientific and technical effort, and eventually the setting of a Technical Advisory Group (TAG) to deepen and revise the proposal made in the book.

For this reason, the effort could be reasonably exploited even for a broader scope matching the *New European Bauhaus Strategy*, striving to achieve a more beautiful, sustainable and inclusive forms of living together. To this end the PA category can be relevant for the industrial asset, which would benefit from the development of a scheme specifically dedicated to it, but also as new topic in general, suitable for many other building types and functions. As a matter of fact, the Italian landscape is unique and attractive right for its natural and cultural values, which are not necessarily found in historic centre or natural protected areas only, but in ordinary scenery, as the Author stressed on.

The proposal described in the book goes right in this direction of protecting/enhancing the value of "everyday" landscapes, while boosting green constructions. And even if I am referring to the Italian market, since I am writing on behalf of GBC Italia, the idea has a great potential to be out scaled in other countries and contexts on the same path of innovation initiated by the Italian green building community represented by our association.

Besides, the proposal goes beyond the typical structure of GBRS as assessment tools: by matching each impact which is measured with a series of possible tactics and practices that can help the company/designer to reduce it, the toolkit becomes more a design-support tool than a mere rating. This is driven by the concept that even if awareness around the impact of factories is growing in general, companies often do not know how and from where to start to mitigate it. In the end, what "Designing sustainable factories" entails is a change of perspective of all the stakeholders in both the field of construction and industry, by offering a practical mean to design a more sustainable built environment through a co-design process and an inclusive approach.

Marco Mari
GBC Italia, President

# Annex 1: Case Studies Short Cards

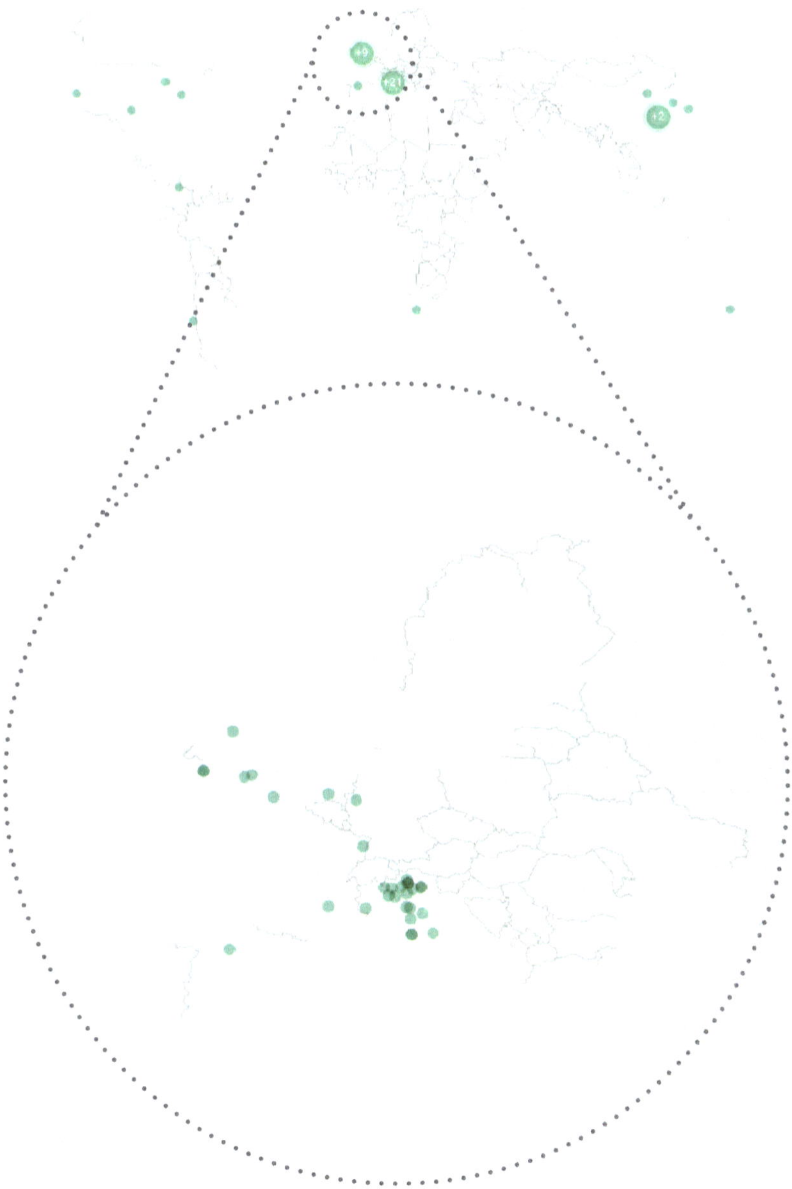


**Geographic distribution of case studies.** World (above) and European (below) scales [created with Datawrapper by the Author].



L. Marchi, *Designing Sustainable Factories*, Advances in Global Change Research 72,
https://doi.org/10.1007/978-3-030-92227-6

| A1—MELINDA | |
|---|---|
| Location | Denno (IT) |
| Date | 2014 |
| Designers | Studio Azzali |
| Client | Melinda |
| Intervention | NC + R |
| Area | 10.000 $m^2$ |
| Reference | www.studioazzali.it |
| *A2—BIO PASTIFICIO IRIS* | |
| Location | Casteldidone (IT) |
| Date | 201 |
| Designers | Architetto Mario Filocca |
| Client | IRIS and Astra Bio |
| Intervention | NC |
| Area | 14.724 $m^2$ |
| Reference | www.irisbio.com |
| *A3—ALMAZARA OLISUR* | |
| Location | O'Higgins (CL) |
| Date | 2009 |
| Designers | GH + A—Guillermo Hevia Architects |
| Client | Olisure S.A |
| Intervention | NC |
| Area | 2.800 $m^2$ |
| Reference | www.guillermohevia.cl |
| *A4—STABILIMENTO MAINA* | |
| Location | Fossano (IT) |
| Date | 2015 |
| Designers | Architetto Gianni Arnaudo |
| Client | Maina Panettoni Spa |
| Intervention | NC + R |
| Area | N.D |
| Reference | www.gianniarnaudo.com |
| *A5—GRANDI SALUMIFICI ITALIANI* | |
| Location | Paganino (IT) |
| Date | 2013 |
| Designers | Politecnica Ingegneria e Architettura |
| Client | Grandi Salumifici Italiani Spa |
| Intervention | R |
| Area | N.D |
| Reference | www.politecnica.it |

(continued)

(continued)

| *A6—SOLARE MANUFAKTUR PETER* | |
|---|---|
| Location | Essen (DE) |
| Date | 2012 |
| Designers | RMA Architekten |
| Client | Peter Backwaren OHG |
| Intervention | NC |
| Area | 2.100 $m^2$ |
| Reference | r-m-a-architekten.de |
| *A7—KERRYGOLD* | |
| Location | Leek (UK) |
| Date | 2008 |
| Designers | Woods Hardwick |
| Client | Kerrygold Ltd |
| Intervention | NC |
| Area | 15.000 $m^2$ |
| Reference | www.woodshardwick.com |
| *A8—PASTIFICIO FELICETTI* | |
| Location | Predazzo (IT) |
| Date | 2014 |
| Designers | RWA - Ruffo Wolf Architetti |
| Client | Monograno Felicetti |
| Intervention | NC |
| Area | 6.000 $m^2$ |
| Reference | www.rwarchitetti.it |
| *A9—CELLE IPOGEE MELINDA* | |
| Location | Tuenetto (IT) |
| Date | 2014 |
| Designers | Tassullo Materiali and research centers |
| Client | Consorzio Melinda |
| Intervention | R |
| Area | 4.000 $m^2$ |
| Reference | www.melinda.it |
| *A10—PELLINI CAFFE'* | |
| Location | Bussolengo (IT) |
| Date | 2006 |
| Designers | Barbi Arca Studio |
| Client | Pellini Caffè |
| Intervention | NC |
| Area | N.D |
| Reference | www.barbiarcastudio.it |

(continued)

(continued)

| *A11—AMADORI* | |
|---|---|
| Location | Cesena (IT) |
| Date | vv |
| Designers | vv |
| Client | Amadori Group |
| Intervention | R |
| Area | N.D |
| Reference | www.amadori.it |
| *A12—BOVETTI* | |
| Location | Lavilledieu (FR) |
| Date | 2010 |
| Designers | CA/PA Architectes |
| Client | Bovetti Chocolats |
| Intervention | NC + R |
| Area | 2.641 $m^2$ |
| Reference | www.capa-arch.eu |
| *A13—C.O.CE.A* | |
| Location | Segno (IT) |
| Date | 2015 |
| Designers | Studio Azzali |
| Client | Consorzio Melinda |
| Intervention | NC |
| Area | 4.800 $m^2$ |
| Reference | www.studioazzali.it |
| *A14—TOFU FACTORY* | |
| Location | Lishui (CH) |
| Date | 2018 |
| Designers | DnA_Design and Architecture |
| Client | Economic Coop. Limited of Caizhai |
| Intervention | NC |
| Area | 1.210 $m^2$ |
| Reference | www.designandarchitecture.net |
| *A15—BROWN SUGAR FACTORY* | |
| Location | Lishui (CH) |
| Date | 2016 |
| Designers | DnA_Design and Architecture |
| Client | Government of Zhangxi Village |
| Intervention | NC |
| Area | 1.230 $m^2$ |
| Reference | www.designandarchitecture.net |

(continued)

(continued)

| | |
|---|---|
| *A16—MARS PETCARE* | |
| Location | Bathurst (AU) |
| Date | 2015 |
| Designers | N.D |
| Client | Mars Petcare |
| Intervention | NC |
| Area | 11.295 $m^2$ |
| Reference | www.usgbc.org |
| *A17—FRITO-LAY* | |
| Location | Plano (TX) |
| Date | 2011 |
| Designers | N.D |
| Client | PepsiCo Frito-Lay |
| Intervention | NC |
| Area | 52.800 $m^2$ |
| Reference | www.usgbc.org |
| *A18—IROHA VILLAGE FACTORY* | |
| Location | Hiroshima (J) |
| Date | 2018 |
| Designers | Hiroshi Nakamura & NAP |
| Client | Fujiiya |
| Intervention | NC |
| Area | 2.862 $m^2$ |
| Reference | www.nakam.info/en/ |
| *A19—OPIFICIO SALPI* | |
| Location | Preci (IT) |
| Date | 2010 |
| Designers | Nothing Studio (Enzo Eusebi + Partners) |
| Client | SAL.PI. Uno srl |
| Intervention | NC |
| Area | 6.200 $m^2$ |
| Reference | www.enzoeusebi.it |
| *A20—HARIM PET FOOD* | |
| Location | Gongju-Si (KR) |
| Date | 2016 |
| Designers | The Beck Group |
| Client | Harim group |
| Intervention | NC |
| Area | 15.900 $m^2$ |
| Reference | www.beckgroup.com |

(continued)

(continued)

| *A21—VAN GELDER* | |
|---|---|
| Location | Ridderkerk (NL) |
| Date | 2019 |
| Designers | Kraaijvanger Architects |
| Client | Van Gelder Ridderkerk B.V |
| Intervention | NC |
| Area | 29.111 $m^2$ |
| Reference | www.kraaijvanger.nl |
| *A22—CAFFÈ MILANI* | |
| Location | Lipomo (I) |
| Date | 2017 |
| Designers | Studio Castiglioni e Nardi Associati |
| Client | Milani Spa |
| Intervention | NC |
| Area | 7.938 $m^2$ |
| Reference | www.studiocastiglioninardi.it |
| *A23—CARNITALIA* | |
| Location | Ospedaletto Lodigiano (IT) |
| Date | 2016 |
| Designers | Studio Castiglioni e Nardi Associati |
| Client | Carnitalia Srl |
| Intervention | NC |
| Area | 11.240 $m^2$ |
| Reference | www.studiocastiglioninardi.it |
| *A24—BELL & EVANS* | |
| Location | Fredericksburg (US) |
| Date | 2021 |
| Designers | Stellar |
| Client | Bell & Evans |
| Intervention | NC |
| Area | 38.230 $m^2$ |
| Reference | www.stellar.net |
| *A25—FELSINEO VEG* | |
| Location | Zola Predosa (IT) |
| Date | 2017 |
| Designers | SICArchitetture |
| Client | FMV Srl |
| Intervention | R |
| Area | 2.485 $m^2$ |
| Reference | sicarchitetture.it |

(continued)

(continued)

| A26—FRATELLI BERETTA | |
|---|---|
| Location | Mount Olive Township (US) |
| Date | 2016 |
| Designers | Marco Castelletti |
| Client | Fratelli Beretta USA |
| Intervention | NC |
| Area | 20.000 $m^2$ |
| Reference | www.marcocastelletti.it |
| *A27—TANGSHAN ORGANIC* | |
| Location | Tangshan (CH) |
| Date | 2016 |
| Designers | Archstudio |
| Client | N.D |
| Intervention | NC |
| Area | 1.720 $m^2$ |
| Reference | www.archstudio.cn |
| *A28—MULINO AGRI '90* | |
| Location | Storo (IT) |
| Date | 2011 |
| Designers | Nexus! Associati |
| Client | Cooperativa Agricola Agri'90 |
| Intervention | NC |
| Area | N.D |
| Reference | www.nexusassociati.com |
| *B1—UC DAVIS WINE BREWERY* | |
| Location | Davis (CA) |
| Date | 2013 |
| Designers | Siegel & Strain Architects |
| Client | August A. Busch III Science Lab |
| Intervention | NC |
| Area | 3.158 $m^2$ |
| Reference | sustainability.ucdavis.edu |
| *B2—CANTINA PIZZOLATO* | |
| Location | Villorba (IT) |
| Date | 2016 |
| Designers | MADE Associati |
| Client | Cantina Pizzolato |
| Intervention | NC + R |
| Area | N.D |
| Reference | www.madeassociati.it |

(continued)

(continued)

| | |
|---|---|
| *B3—CANTINA ANTINORI* | |
| Location | San Casciano, Val di Pesa (IT) |
| Date | 2013 |
| Designers | Archea Associati |
| Client | Marchesi Antinori Srl |
| Intervention | NC |
| Area | 49.000 $m^2$ |
| Reference | www.archea.it |
| *B4—BODEGAS PORTIA* | |
| Location | Gumiel de Izán (ES) |
| Date | 2010 |
| Designers | Foster and partners |
| Client | Faustino Group |
| Intervention | NC |
| Area | 12.500 $m^2$ |
| Reference | www.fosterandpartners.com |
| *B5—MACALLAN DISTILLERY* | |
| Location | Speyside (UK) |
| Date | 2018 |
| Designers | Rogers Stirk Harbour + Partners |
| Client | The Edrington Group |
| Intervention | NC |
| Area | 14.800 $m^2$ |
| Reference | www.rsh-p.com |
| *B6—DALMUNACH DISTILLERY* | |
| Location | Carron (UK) |
| Date | 2015 |
| Designers | Archial NORR |
| Client | Chivas Brothers Ltd |
| Intervention | NC |
| Area | 4.500 $m^2$ |
| Reference | norr.com |
| *B7—CARLSBERG* | |
| Location | Obernai (FR) |
| Date | 2014 |
| Designers | S&AA |
| Client | Carlsberg Group |
| Intervention | NC |
| Area | 9.626 $m^2$ |
| Reference | www.schweitzer-associes.com |

(continued)

(continued)

| *B8—SAN PELLEGRINO* | |
|---|---|
| Location | San Pellegrino Terme (IT) |
| Date | ongoing |
| Designers | BIG—Bjarke Ingels Group |
| Client | San Pellegrino |
| Intervention | NC |
| Area | 17.500 $m^2$ |
| Reference | www.big.dk |
| *B9—CANTINA DI COLLEMASSARI* | |
| Location | Cinigiano (IT) |
| Date | 2005 |
| Designers | Edoardo Milesi and Archos |
| Client | Collemassari |
| Intervention | NC |
| Area | N.D |
| Reference | www.archos.it |
| *B10—CANTINA PIEVE VECCHIA* | |
| Location | Campagnatico (IT) |
| Date | 2012 |
| Designers | Cini Boeri |
| Client | Collemassari |
| Intervention | NC |
| Area | N.D |
| Reference | www.ciniboeriarchitetti.com |
| *B11—VALPRÈ* | |
| Location | Heidelberg (ZA) |
| Date | 2011 |
| Designers | Luis Praveres |
| Client | Valprè Spring Water |
| Intervention | NC |
| Area | 15.607 $m^2$ |
| Reference | www.usgbc.org |
| *B12—BAVARIA BREWERY* | |
| Location | Bucaramanga (CO) |
| Date | 2011 |
| Designers | Construcciones Planificadas |
| Client | Bavaria S.A |
| Intervention | NC |
| Area | 25.167 $m^2$ |
| Reference | www.construccionesplanificadas.com |

(continued)

(continued)

| *B13—ST. JAMES GATE BREWERY* | |
|---|---|
| Location | Dublin (UK) |
| Date | 2011 |
| Designers | RKD Achitects |
| Client | Diageo Ireland |
| Intervention | NC |
| Area | 9.315 $m^2$ |
| Reference | www.diageo.com |
| *B14—BOMBAY SAPPHIRE DISTILLERY* | |
| Location | Laverstoke Mill, Whitchurch (UK) |
| Date | 2014 |
| Designers | Heatherwick Studio; GWP Architecture |
| Client | Bombay Spirits Ltd |
| Intervention | R |
| Area | 2.500 $m^2$ |
| Reference | gwp-arch.com |
| *B15—CE.VI.V* | |
| Location | Susegana (IT) |
| Date | 2018 |
| Designers | ReisArchitettura |
| Client | CE.VI.V. Srl |
| Intervention | NC |
| Area | 2.230 $m^2$ |
| Reference | reisarchitettura.it |
| *B16—ROE & CO DISTILLERY* | |
| Location | Dublin (UK) |
| Date | 2019 |
| Designers | RKD Achitects |
| Client | Roe & Co |
| Intervention | R |
| Area | 3.500 $m^2$ |
| Reference | www.roeandcowhiskey.com |
| *B17—DISTILLERIA ZANIN* | |
| Location | Zugliano (IT) |
| Date | 2013 |
| Designers | Roberto Munaretto |
| Client | Distilleria Zanin Srl |
| Intervention | NC + R |
| Area | 25.000 $m^2$ |
| Reference | www.zanin.it |

# Annex 2: Library of Tactics


L. Marchi, *Designing Sustainable Factories*, Advances in Global Change Research 72,
https://doi.org/10.1007/978-3-030-92227-6

## Synoptic Circle

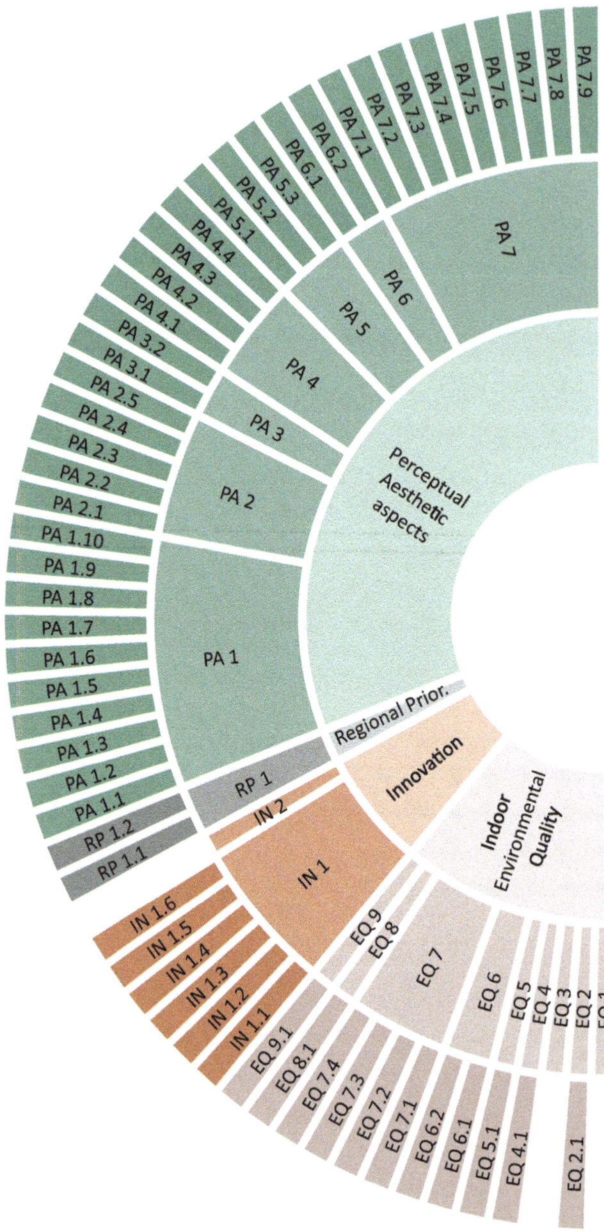

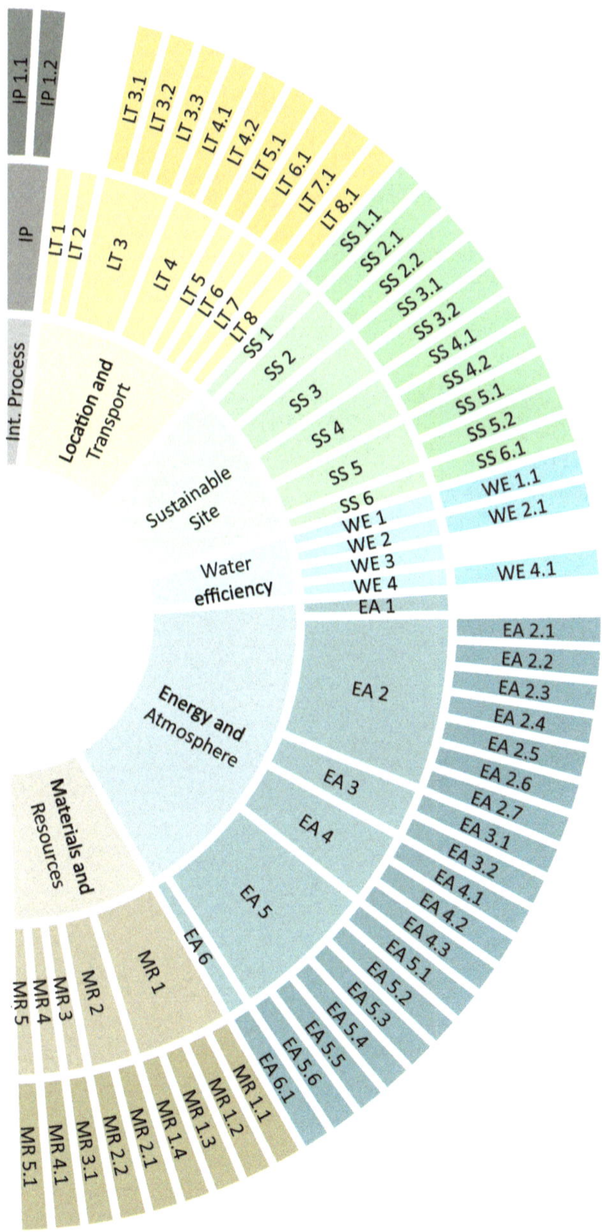
Int. Process
IP
IP 1.1
IP 1.2
Location and Transport
LT 1
LT 2
LT 3
LT 4
LT 5
LT 6
LT 7
LT 8
LT 3.1
LT 3.2
LT 3.3
LT 4.1
LT 4.2
LT 5.1
LT 6.1
LT 7.1
LT 8.1
Sustainable Site
SS 1
SS 2
SS 3
SS 4
SS 5
SS 6
SS 1.1
SS 2.1
SS 2.2
SS 3.1
SS 3.2
SS 4.1
SS 4.2
SS 5.1
SS 5.2
SS 6.1
Water efficiency
WE 1
WE 2
WE 3
WE 4
WE 1.1
WE 2.1
WE 4.1
Energy and Atmosphere
EA 1
EA 2
EA 3
EA 4
EA 5
EA 6
EA 2.1
EA 2.2
EA 2.3
EA 2.4
EA 2.5
EA 2.6
EA 2.7
EA 3.1
EA 3.2
EA 4.1
EA 4.2
EA 4.3
EA 5.1
EA 5.2
EA 5.3
EA 5.4
EA 5.5
EA 5.6
EA 6.1
Materials and Resources
MR 1
MR 2
MR 3
MR 4
MR 5
MR 1.1
MR 1.2
MR 1.3
MR 1.4
MR 2.1
MR 2.2
MR 3.1
MR 4.1
MR 5.1

## Tactics list

IP 1.1 Stakeholders engagement
IP 1.2 Multidisciplinary design team

LT 3.1 Building extension
LT 3.2 Recover of a degraded area
LT 3.3 Adaptive reuse
LT 4.1 Centralisation of production and logistics
LT 4.2 Urban location
LT 5.1 Connection with public transport
LT 6.1 Bike rack and showers
LT 7.1 Limited car parks
LT 8.1 Green vehicles

SS 1.1 Site ventilation and cooling
SS 2.1 Flora and fauna protection
SS 2.2 Ecological networks
SS 3.1 Green spaces design
SS 3.2 Open space design linked with building layout
SS 4.1 Rain water collection
SS 4.2 Pervious surfaces
SS 5.1 Green roof
SS 5.2 Green folded surfaces
SS 6.1 Light pollution prevention

WE 1.1 Outdoor reuse of rainwater
WE 2.1 Water saving technologies
WE 4.1 Water management systems

EA 2.1 Ventilated façade
EA 2.2 Shading systems
EA 2.3 Thermal insulation
EA 2.4 Thermal mass
EA 2.5 Ground thermal proprieties
EA 2.6 Green façade
**EA 2.7 Devices for efficient heating**
EA 3.1 Timed or men-presence activation systems
EA 3.2 Advanced energy management
EA 4.1 Local energy provider
EA 4.2 On-site energy production
EA 4.3 Scalable technical building systems
EA 5.1 Solar energy

EA 5.2 Geothermal energy
EA 5.3 Energy from biomass
EA 5.4 Cogeneration
EA 5.5 Wind energy
**EA 5.6 Off-site green energy production or purchase**
EA 6.1 GHG free refrigeration

MR 1.1 Recyclable structure
MR 1.2 Modular structure and/or components
**MR 1.3 Building retrofitting**
MR 1.4 Durability and low maintenance
MR 2.1 Low life cycle impact
MR 2.2 Recycled/recyclable materials
MR 3.1 Local materials
MR 4.1 Natural materials
MR 5.1 Recover of construction and demolition waste

**EQ 2.1 Eco-friendly finishing**
EQ 5.1 Natural ventilation
EQ 6.1 Smart lighting
**EQ 6.2 Highly-efficient lamps**
**EQ 7.1 Diffused top lighting**
EQ 7.2 Glass façade
EQ 7.3 Filter façade
**EQ 7.4 High reflectance indoor finishing**
EQ 8.1 View of the scenery
EQ 9.1 Protection from noise of machinery

IN 1.1 Process automation
IN 1.2 Recover of heat/energy from processes
IN 1.3 Energy from waste streams
IN 1.4 Treatment and reuse of grey water
IN 1.5 Treatment and reuse of process water
IN 1.6 Technology to abate pollution

RP 1.1 Sharing of facilities in consortium
RP 1.2 Sharing of services and mixed use

PA 1.1 Visual screen
PA 1.2 Horizontal dimension
PA 1.3 Landform

PA 1.4 Underground building
PA 1.5 Reference to local constructions
PA 1.6 Reference to the skyline
PA 1.7 Volume breakdown
PA 1.8 Clear layout
PA 1.9 Kempt borders
PA 1.10 Hidden setting
PA 2.1 Colour from the context
PA 2.2 Pattern from the context
PA 2.3 Vegetated roof
PA 2.4 Materials from the context
PA 2.5 Graphic design
PA 3.1 Terrain or vegetation barrier
PA 3.2 Technological barrier
PA 4.1 Multifunction facility
PA 4.2 Territorial promotion and Social Responsibility
PA 4.3 Public park
PA 4.4 Visibility and identity of open space
PA 5.1 Services for workers
PA 5.2 Ergonomic work place
PA 5.3 Flexible space
PA 6.1 Consideration of the landscape's value
PA 6.2 Perspective and visual channels
PA 7.1 Consistency with branding
PA 7.2 Reference to product's ingredients
PA 7.3 Showcase of technology
PA 7.4 Building as showcase of the product
PA 7.5 Consistency with company values
**PA 7.6 Green Building Certification**
PA 7.7 Experiences for consumers
PA 7.8 Dynamic image on the road
PA 7.9 Process communication outwards

## How to read the tactic card

Tactic's code and title

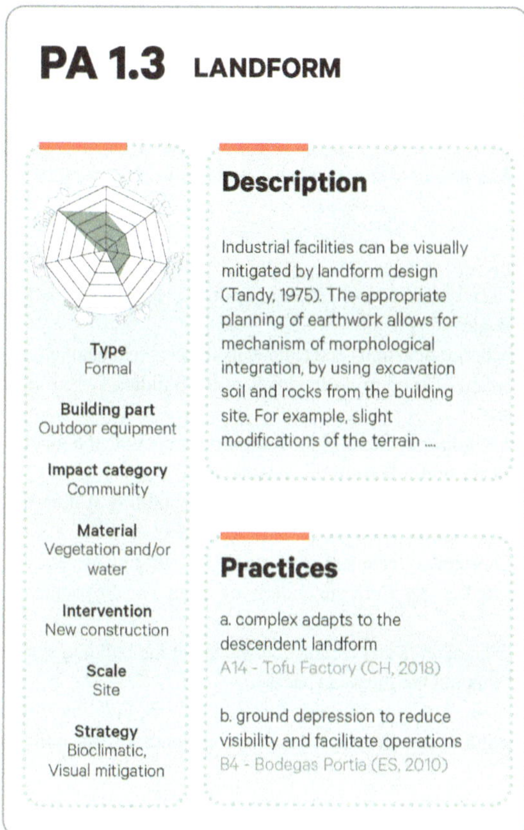


General description with examples and references

List of good practices and related case studies (in grey)

Impact category radar and faceted classification

## Glossary

**Type**

- **Technological**: refers to one or more building technological units (ref. UNI 8290).
- **Formal**: concerns the composition of volumes, their shape, the relationship between built and void spaces, the quality of the resulting spaces; or the use of elements of the construction or trick that modify the building's visual perception.
- **Functional**: concerns a service offered to possible users (e.g., leisure function).
- **Management:** refers to approaches adopted in the design or operational phase to manage the relation with workers, consumers, inhabitants, or other stakeholders relevant to the specific case; or in a broader sense a design choice that has a strategic value (e.g., zero soil consumption).

**Building part**

Retrieved by the Italian standard UNI 8290-1:1981 + A122:1983 'Edilizia residenziale. Sistema tecnologico. Classificazione e terminologia'

- **Load-bearing structure**: set of technological units and technical elements that belong to the building system and have the function of supporting the loads of the building system itself and of statically connecting its parts.
- **Vertical envelope**: set of vertical technical elements of the building system with the function of separating the internal spaces of the building from the outside.
- **Top horizontal envelope**: set of horizontal technical elements of the building system with the function of separating the internal spaces from the external space above.
- **Low horizontal envelope**: set of horizontal technical elements of the building system with the function of separating the internal spaces from the underlying ground or the foundation structures.
- **Internal partition**: set of technological units and technical elements of the building system with the function of dividing and shaping the internal spaces.
- **External partition**: set of technological units and technical elements of the building system with the function of dividing and conforming the external spaces connected with the building system itself.
- **Systems**: set of technological units and technical elements of the building system with the function of allowing the use of energy flows, information and materials required by users and the consequent removal of any waste.
- **Indoor equipment**: set of technological units and technical elements connected to the building system having the function of allowing or facilitating the user activities in the internal spaces of the building system itself.
- **Outdoor equipment**: set of technological units and technical elements having the function of allowing or facilitating the user activities in outdoor spaces.
- **N.D. (Not Defined)**: diverse technical units are seamlessly affected.
- **N.A. (Not Applicable)**: the tactics do not affect the building physical elements.

**Impact category**

- **Climate change**: to reverse contribution to global climate change (e.g., GHG emissions reduction from building operations energy use).
- **Health and Well-being**: to enhance individual human health and well-being (e.g., support occupant comfort and wellness).
- **Water resource**: to protect and restore water resources (e.g., water conservation).
- **Biodiversity**: to protect, enhance and restore biodiversity and ecosystem services (e.g., local biodiversity, habitat protection and open spaces).
- **Material resource**: to promote sustainable and regenerative material resources cycles (e.g., reduce raw material resource extraction).
- **Green economy**: to build a greener economy (e.g., enhance the value proposition of green building).
- **Community**: to enhance social equity, environmental justice, and community quality of life (e.g., create a strong sense of place).

**Materials**

- **Wood; Concrete; Metals; Glass; Coating and paint; Vegetation and/or water**
- **N.D. (Not Defined)**: several materials can be used to the scope.
- **N.A. (Not Applicable)**: the tactic is not directly linked with a construction material.

**Intervention**

- **New construction**: includes all projects that deal with new construction of buildings/ infrastructures/facilities; deep urban regeneration; deep renovation with different footprint; increased volume or expansion.
- **Renovation**: includes building renovation within footprint, and ordinary or extraordinary maintenance (e.g., energy retrofitting).
- **N.D. (Not Defined)**: the tactic can be applied either to New construction or Renovation.

**Scale**

- **Site**: refers to the relation of the building with the context, in terms of infrastructures, open spaces, natural environment, site perception, services offered to the community.
- **Building**: refers to the architecture of the factory impacting in various ways the building technological units (e.g., envelope, load-bearing structure).
- **Process**: refers to the integration of the building technical systems with the process, in terms of resource use and environmental impacts on the local ecosystem.
- **N.A. (Not Applicable)**: the tactic cannot be linked directly to a design scale.

**Strategy**

The category classifies tactics per physical phenomenon involved, per service offered, or more in general per design concept.

- **Energy efficiency**: helps to reduce the building's and/or process's energy consumption, or to recover some of the residual energy from the building operations.
- **Renewable resources**: the energy source and/or materials used for the construction are renewable.

- **New material saving**: materials used in the facility construction are recovered, recycled, or designed to last for a long time before being recycled or recovered.
- **Water saving**: refers to water-saving technologies or re-use of water from operations to reduce water demand for running the building/process.
- **Bioclimatic**: limits the use of mechanical systems for outdoor and indoor comfort while increasing positive thermal exchange between the building and the surrounding environment.
- **Ecosystem protection**: protects local flora and fauna, as well as pre-existing natural balance, and in certain cases strengthens them by establishing new links (e.g., by nature-based solutions).
- **Automatic system**: implements Information Technology to optimise the use of resources by both the building technical systems and the process.
- **Soil saving**: reduces the extent of new hardscapes and/or of the building footprint; alternatively, reuses soils that are already compromised (e.g., brownfield).
- **Air pollution reduction**: uses materials or technology that limit or even reduce greenhouse gas emissions and other hazardous pollutants in the atmosphere.
- **Sustainable mobility**: promotes soft mobility, such as the use of public or shared transportation, as well as fuel-efficient or non-fossil-fuel cars.

- **Community engagement**: engages local residents and other stakeholders (e.g., businesses, associations) in the design process and/or provides them with activities and services that enrich the site functional offer.
- **Flexibility**: spaces and structures are designed to adapt across time, whether it is a day or the building whole life cycle (e.g., movable partitions, large multi-purpose spaces).
- **Visual comfort**: favours the users' visual and light comfort through technology or materials that provide adequate level of daylight or at least use the temperature of natural light.
- **Acoustic comfort**: reduces the propagation of sound generated by machineries inside or outside the building by means of technology or room features or, vice versa, limits noise pollution from the context (e.g., from congested roads).
- **Well-being of workers**: promotes the well-being and health of workers by designing high quality space and services for both work and restoration.

- **Visual mitigation**: reduces the visibility of the building in the scenery through composition, chromatic and graphic design practices.
- **Brand image**: emphasizes the presence of the factory in the landscape without detracting quality, but impressing the observant with a memorable image.
- **Physical link with the context**: connects the site with the context by lines and symbols that recall the scenery, its colours and materials.
- **Green marketing**: represents a marketing means for the firm that communicate its excellent environmental performances, but also aesthetically balanced and socially respectful practices.

## Tactics' cards

# IP 1.1 STAKEHOLDERS ENGAGEMENT

**Type**
Management

**Building part**
N.A.

**Impact category**
Community

**Material**
N.A.

**Intervention**
N.D.

**Scale**
N.A.

**Strategy**
Community engagement

**Description**

From the very beginning of the project, involve the many and diverse people affected by the transformation (i.e., stakeholders) in order to acquire valuable information about the site, both tangible and intangible.

This might also facilitate the design team to obtain the support and approval of the community settled in and nearby the selected area. In fact, the ability to set a dialogue with the context and the subjects that animate is critical to the success and sustainability of a project.

Engaging tools that are useful to this purpose are for instance participatory meetings and informal walks, as well as design charrettes between local associations, residents, public administration, professionals, and the company. Despite initially appearing to be time-consuming, this approach often turns out to be time and money-saving throughout the entire design stage and building life cycle.

**Practices**

a) involvement of local environmental associations to designing ecosystem protection strategies
B6 - Dalmunach distillery (UK, 2015)

b) active participation of citizens, associations and municipality in the design and construction stages
B14 - Bombay Sapphire Distillery (UK, 2013)

# IP 1.2 MULTIDISCIPLINARY DESIGN TEAM

**Type**
Management

**Building part**
N.A.

**Impact category**
Climate change

**Material**
N.A.

**Intervention**
N.D.

**Scale**
N.A.

**Strategy**
-

**Description**

Enhance the cooperation among various professionals for the management of complex projects such as the design, construction, and maintenance of an industrial building.

A good project must consider the socioeconomic, urban, functional, environmental, and landscape aspects of the location in an integrated manner: to do so, it is critical that professionals from diverse backgrounds collaborate on the proposal in synergy (Busquet I Fabregas, 2007).

In fact, a multidisciplinary design team work aids in minimizing negative impacts on all fronts while aiming for the overall project excellence. The integrative design process is particularly helpful to this regard (U.S. GBC, 2014).

Building Information Modelling (BIM) is also a useful design tool that facilitates and structures the integration of the various disciplines and skills involved.

**Practices**

a) multidisciplinary team

A6 - Solare Manufaktur Peter (DE, 2012)

B6 - Dalmunach distillery (UK, 2015)

B14 - Bombay Sapphire Distillery (UK, 2013)

# LT 3.1 BUILDING EXTENSION

**Type**
Formal

**Building part**
N.D.

**Impact category**
Material resources

**Material**
N.D.

**Intervention**
New construction

**Scale**
Site

**Strategy**
Soil saving

**Description**

The expansion of production or a diverse organization of the manufacturing process often requires new spaces, when not a new complex.

Taking advantage of the company's already built area allows it to restrict the intervention environmental and landscape detrimental effects: not only is the consumption of new soil minimized, but the potential consequences related to freight logistics and transfer to other places reduce.

If the conditions allow, it is preferable to maintain continuity with the existing building to integrate the processes and avoid penalizing production disruptions.

**Practices**

a) quick intervention not to interrupt manufacturing
A1 - Melinda (IT, 2014)
A7 - Kerrygold (UK, 2008)

b) new construction is seamless with the existing building
A4 - Stabilimento Maina (IT, 2015)
A12 - Bovetti (FR, 2010)
A16 - Mars Petcare (AU, 2015)
A22 - Caffè Milani (IT, 2017)
A28 - Mulino Agri'90 (IT, 2011)
B2 - Cantina Pizzolato (IT, 2016)
B7 - Carlsberg (FR, 2014)
B12 - Bavaria Brewery (CO, 2011)
B15 - CE.VI.V. (IT, 2018)

# LT 3.2 RECOVER OF A DEGRADED AREA

**Type**
Technological

**Building part**
N.D.

**Impact category**
Biodiversity

**Material**
N.D.

**Intervention**
N.D.

**Scale**
Site

**Strategy**
Soil saving,
Green marketing

**Description**

In line with the global challenge of reducing soil consumption, combating the under-use of areas already designated for production activities (O.P.U.Q.P., 2015) entails redeveloping abandoned buildings, plants, and infrastructure.

**The redevelopment of a site in a brownfield - a land where** the presence or potential presence of hazardous substances, pollutants, or contaminants can cause problems - or in a degraded or abandoned area can start a virtuous cycle of regeneration for the entire location.

As a pilot case, a successful sustainable factory might **demonstrate to "neighbours" the benefits of renovation** and possibly increase the value of the area, making it more appealing to potential investors.

**Practices**

a) pilot project to renovate the whole area
A8 - Pastificio Felicetti (IT, 2014)

**b) buildings with energy efficiency improvement**
A17- Frito-Lay (TX, 2011)

c) reclaimed land without existing constructions
A18 - Iroha Village Factory (J, 2018)

d) building within the former construction's footprint
A23 - Carnitalia (IT, 2016)
B6 - Dalmunach distillery (UK, 2015)

# LT 3.3 ADAPTIVE REUSE

**Type**
Technological

**Building part**
N.D.

**Impact category**
Material resources

**Material**
N.D.

**Intervention**
Renovation

**Scale**
Building

**Strategy**
New resource saving,
Green marketing

**Description**

Limit consumption of new soil and the environmental impact of demolition and reconstruction by reusing a building originally intended for another purpose.
In this regard, production requirements might be a limitation, but certain examples of adaptive reuse indicate that, where suitable, this results in significant environmental and economic benefits.
The reuse prerogatives include environmental safety and landscape protection (O.P.U.Q.P., 2015): for example, if the area needs to be clean there may be restrictions that make the reuse process difficult.

**Practices**

a) reuse of an underground quarry for food conservation
A9 - Celle ipogee Melinda (IT, 2004)

b) reuse of an existing industrial building for a new productive function
B14 - Bombay Sapphire Distillery (UK, 2013)
B16 - Roe & Co Distillery (IE, 2019)
B17 - Distilleria Zanin (IT, 2013)

# LT 4.1 CENTRALISATION OF PRODUCTION & LOGISTICS

**Type**
Management

**Building part**
N.A.

**Impact category**
Climate change

**Material**
N.A.

**Intervention**
New construction

**Scale**
Site

**Strategy**
Soil saving,
Sustainable mobility

**Description**

Concentrating production in a single place might significantly reduce the economic and environmental costs associated with goods transport (particularly semi-finished products) and thus heavy traffic.

This strategy is especially effective for small and medium-sized businesses with a few production sites: in some cases, it can also be convenient to locate the factory close to the reference market.

**Practices**

a) local concentration of production to reduce heavy traffic

A8 - Pastificio Felicetti (IT, 2014)

A24 - Bell & Evans (U.S., ongoing)

b) location near to the primary market

B11 - Valprè (ZA, 2011)

# LT 4.2 URBAN LOCATION

**Type**
Management

**Building part**
N.D.

**Impact category**
Climate change

**Material**
N.A.

**Intervention**
New construction

**Scale**
Site

**Strategy**
Soil saving

**Description**

The selection of a location is the most "far-reaching" decision a businessman has to take when planning to build a new factory (Reichardt, 2015).

Today many recommend re-establishing the relationship between factory and city. Traditionally the factory is a separate place from the city and people's living spaces for reasons of hygiene, safety, and, for a long time, economic reasons linked to the cost of land and proximity to raw material procurement.

Some virtuous experiences of workers' villages and industrial cities represented a counter-trend in the XX century, to which today's strategies of urban localization of industry seem to refer.

Now the integration of factories into residential areas appears to be a promising future solution (Hermann et al., 2015), based on the assumption that industry's impact on the environment is reducing and it can provide good access to services such as shops, public transport, kindergarten and others (U.S. GBC, 2014).

An urban location makes it easier to connect the factory to the community by pedestrian paths and bicycle lanes, encouraging people to walk or bike instead of drive, not only helping to reduce air pollution but also promoting physical activity (U.S. GBC, 2014). On the contrary, a building developed in a greenfield whose occupants must drive long distances may contribute to greenhouse gas emissions, as well as the destruction of natural habitat due to the infrastructure development.

**Practices**

a) location allows an easy connection with urban services as shops and bank
B13 - St. James Gate Brewery (UK, 2011)

b) intermediate location between village and countryside as filter element
A15 - Brown Sugar Factory (CH, 2016)

# LT 5.1 CONNECTION WITH PUBLIC TRANSPORT

**Type**
Management

**Building part**
N.A.

**Impact category**
Climate change

**Material**
N.A.

**Intervention**
N.D.

**Scale**
Site

**Strategy**
Atmospheric pollution reduction,
Sustainable mobility

**Description**

Encourage workers to use public transportation.
**The access to efficient public transit is largely dependent** on the location of the manufacturing site: that is, the closer the factory is to an urbanized area, the better the link with urban or suburban transportation lines will likely be.
**If the service is insufficient or inadequate, the company** can encourage sustainable mobility by applying requests for integration to the administration and the transportation company if the numbers allow it, or by arranging a shuttle service for employees.

**Practices**

a) easy access to public transport such as tram, bus or other commuting means
B11 - Valprè (ZA, 2011)
B13 - St. James Gate Brewery (UK, 2011)

# LT 6.1 BIKE RACK AND SHOWERS

**Type**
Functional

**Building part**
Outdoor equipment

**Impact category**
Health and Well-being

**Material**
N.D.

**Intervention**
N.D.

**Scale**
Site

**Strategy**
Air pollution reduction,
Sustainable mobility,
Green marketing

**Description**

Promoting sustainable mobility, and especially bicycle use by employees, entails offering services to those who commute by bike. This in turn means promoting physical activity and as a result improving workers' health (U.S. GBC, 2014).

To this end, many building sustainability certification standards, including LEED, reward the provision of secured or similarly protected bicycle storage facilities, as well as the availability of changing rooms and showers within the company's building (Grant, 1997; O.P.U.Q.P., 2015).

It is also recommended that the company provide a set of bicycles available to its employees.

**Practices**

a) bicycle stands, shower and changing rooms available for workers

B11 - Valprè (ZA, 2011)

B13 - St. James Gate Brewery (UK, 2011)

# LT 7.1 LIMITED CAR PARKS

**Type**
Functional

**Building part**
Outdoor equipment

**Impact category**
Climate change

**Material**
N.D.

**Intervention**
New construction

**Scale**
Site

**Strategy**
Soil saving,
Sustainable mobility

**Description**

Reducing the number or size of car parks can benefit both the firm (e.g., a smaller site required for the construction) and the environment (e.g., limiting the pervious surfaces). This can also spark interest in alternative transportation options (U.S. GBC, 2014).

There are several ways to reduce car parks. The first is to share the area with other facilities and services, as well as for loading-unloading area and storage places. Another is promoting alternative modes of transportation that are less polluting and do not require large parking lots (e.g., bicycles) or providing a collective business transportation service (e.g., shuttles) if the number of workers who could use is significant (Grant, 1997).

Similarly, the O.P.U.Q.P. (2015) suggests reducing and rationalizing the management, storage, and parking areas by exploiting the spaces between the buildings.

Alternatively, if the circumstances allow, it is still a good practice to build underground parking lots that are hidden from view and that, if limited within the footprint of the building, do not increase land consumption.

**Practices**

a) underground park
A1 - Melinda (IT, 2014)

# LT 8.1 GREEN VEHICLES

**Type**
Management

**Building part**
N.A.

**Impact category**
Green economy

**Material**
N.A.

**Intervention**
N.D.

**Scale**
N.A.

**Strategy**
Air pollution reduction,
Sustainable mobility,
Green marketing

**Description**

As transport and logistics cover a major share of the company's environmental impact, including atmospheric and noise pollution, equipping the fleet with sustainable vehicles can help.

Hence, the company fleet (trucks and employee automobiles) can be transformed into more fuel-efficient vehicles or powered entirely by renewable energy sources. In addition, employees can be encouraged to use private electric vehicles by reserving a portion of the parking places for these vehicles or supporting car-sharing, ride-sharing, or car-pooling.

**Practices**

a) arrangement for future electric fleet that will be fed by on-site renewable energy
A6 - Solare Manufaktur Peter (DE, 2012)

b) highly efficient company vehicles
B11 - Valprè (ZA, 2011)

c) dedicated park lots for efficient, green cars or carpooling
B13 - St. James Gate Brewery (UK, 2011)

# SS 1.1 SITE VENTILATION AND COOLING

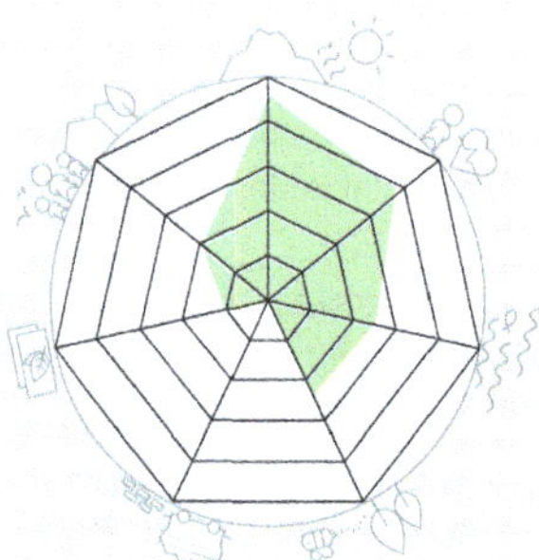

**Type**
Technological

**Building part**
Outdoor equipment

**Impact category**
Climate change

**Material**
Vegetation and/or water

**Intervention**
N.D.

**Scale**
Site

**Strategy**
Bioclimatic,
Well-being of workers

**Description**

Analyse the micro-climatic characteristics of the place to learn how to take advantage of them in the project, as there is no best orientation valid for each context or geographical location.

Therefore, it is important to assess the local conditions and possibly detect the In sustainable design in fact it is critical to identify the direction of winds and breezes in order to design the building layout to promote ventilation and passive cooling. The general concept is that prevailing winds must be shielded with windbreak components in winter and exploited in summer. This can be obtained by means of proper ventilation corridors aligned with the prevailing winds and sea-land breezes; or the positioning of water surfaces so that the air currents are also cooled (O.P.U.Q.P., 2015). Last but not least, the wind analysis is also useful to understand where smokes and polluting particles will be transported, and therefore to insert windbreak elements in correspondence of these directions, respecting the inhabitants of the place.

For the same reason it is recommended orienting the complex to optimize free solar gains and **benefits** of natural ventilation (e.g., transverse or ascending). Similarly, hedges, rows, or pergolas can be used to shade buildings and paths.

Furthermore, by constructing correctly exposed walls on **which water flows or pools of water near to the building,** in the central Europe latitude, can improve the comfort of work environments on hot dry days. Dessì et al. (2018) describe some interesting strategies to improve the outdoor micro-climate by using water (e.g. along the footpath, vaporised).

**Practices**

a) courtyard design to favour natural ventilation
A27 - Tangshan Organic (CH, 2016)

# SS 2.1 FLORA AND FAUNA PROTECTION

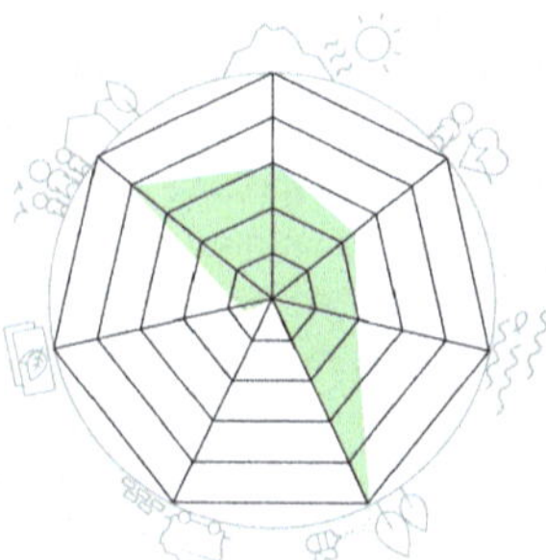

**Type**
Technological

**Building part**
Outdoor equipment

**Impact category**
Biodiversity

**Material**
Vegetation and/or water

**Intervention**
N.D.

**Scale**
Site

**Strategy**
Ecosystem protection,
Physical link with the context

**Description**

Preserve existing ecosystems and local flora and wildlife by limiting environmental fragmentation, or the disruption of pre-existing ecological networks. In general, vegetation borders are preferred to fences since they do not obstruct the passage of wild animals.

It is recommended prioritizing native plant species over purely ornamental ones (O.P.U.Q.P., 2015), or creating new habitats by combining existing species with others that are beneficial to them (e.g. ponds and greenhouses) (Herrmann, 2015). Nature-based solutions are largely used in this field.

Native and adapted plant species are often associated with low maintenance because they thrive without irrigation, pesticides, or fertilizers. Certain plants can even enhance soil nutrients, others naturally deter pests (U.S. GBC, 2014). They also support water efficiency goals because these plants typically do not need artificial irrigation. Xeri-scaping is also helpful: that is use of drought-tolerant native or adapted plants along with rocks, bark mulch, and other landscape elements.

**Practices**

a) wild vegetation in green roof
B5 - Macallan distillery (UK, 2018)

b) use of native plants requiring little water
B11 - Valprè (ZA, 2011)

c) after deviating the river flow for the biodiversity improvement, expert have relocated fishes
B14 - Bombay Sapphire Distillery (UK, 2013)

# SS 2.2 ECOLOGICAL NETWORKS

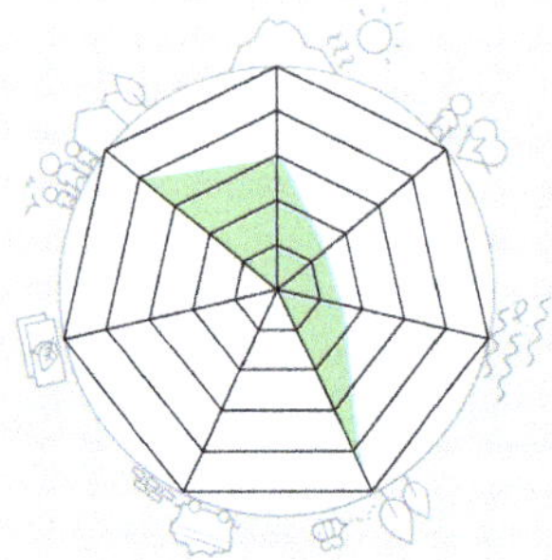

**Type**
Technological

**Building part**
Outdoor equipment

**Impact category**
Biodiversity

**Material**
Vegetation and/or water

**Intervention**
N.D.

**Scale**
Site

**Strategy**
Ecosystem protection,
Physical link with the context

**Description**

Strengthen the link between the site's environmental infrastructures and existing ecological and landscape systems. This can be attained by respecting historical vestiges, urban fabrics, and existing infrastructures, not altering the natural waterways; by locating green areas near forests, green masses, or natural spaces of interest; by developing a network of green areas that operate as visual and biodiversity corridors; by establishing transitional green belts.

Then, arrange the green system to favour continuity with the existing natural environmental matrix, for example, by carefully planting native plant species in open spaces or even on the building itself (e.g. roof, façade). As for the previous tactic, the choice of species to plant must be guided by a deep knowledge of ecological and geographical conditions, and the observation of the plant models already in place must inspire the design (Busquet i Fabregas, 2007). Along the site boundary, prefer living hedges and green slopes to metal or concrete fences (O.P.U.Q.P., 2015). For example "ha-ha" borders might be very helpful to create a vertical green barrier while preserving an uninterrupted view of the landscape beyond.

**Practices**

N.D.

# SS 3.1 GREEN SPACE DESIGN

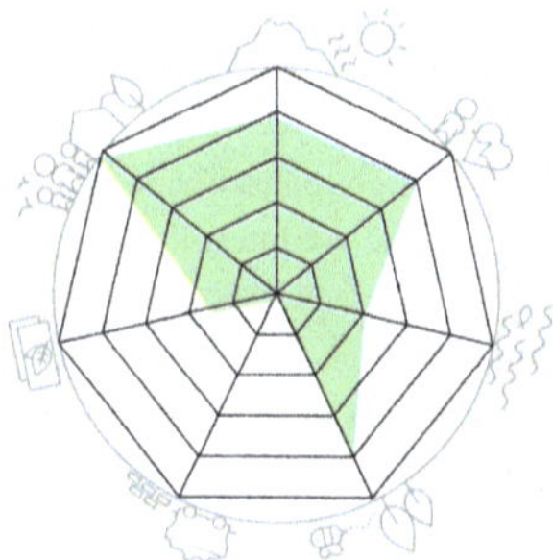

**Type**
Technological

**Building part**
Outdoor equipment

**Impact category**
Community

**Material**
Vegetation and/or water

**Intervention**
N.D.

**Scale**
Site

**Strategy**
Bioclimatic,
Well-being of workers

**Description**

Design green areas to produce both environmental and landscape positive effects. Although they are frequently regarded as simply residual spaces, green areas can greatly contribute to improving the environmental, landscape and social quality of the site (e.g., the well-being of employees and the attractiveness of the firm).
A well-designed green space may help improve the microclimate and cut the site's energy demand too: for instance, by using hedges and rows for protection from strong winds or reverberations; sunshades and pergolas for shading; and water surfaces for passive cooling (Grant, 1997; O.P.U.Q.P., 2015).
Moreover, since they are often as visible as the façade, it is critical to keep green areas tidy and clean; to establish a common plant range and define unitary criteria for structuring different functional areas; to clearly characterize it; and to concentrate built-up spaces and identify representative areas (Busquet i Fabregas, 2007).

**Practices**

a) green areas, water basin or foot/bike paths
A2 - Bio Pastificio IRIS (IT, 2015); A19 - Opificio Salpi (IT, 2010); B5 - Macallan distillery (UK, 2018); B6 - Dalmunach distillery (UK, 2015); B7 - Carlsberg (FR, 2014); B8 - San Pellegrino (IT, ongoing)

b) park for visitors and their animals
A20 - Harim Pet Food (KR, 2016)

d) fruit trees planted in a regular open space
A21 - Van Gelder (NL, 2019)

e) phyto-depuration as design element
B1 - UC Davis Wine Brewery (CA, 2013)

g) reopening of the water canal within the lot and demolition of buildings to create a larger courtyard
B14 - Bombay Sapphire Distillery (UK)

# SS 3.2 OPEN SPACE LINKED WITH BUILDING LAYOUT

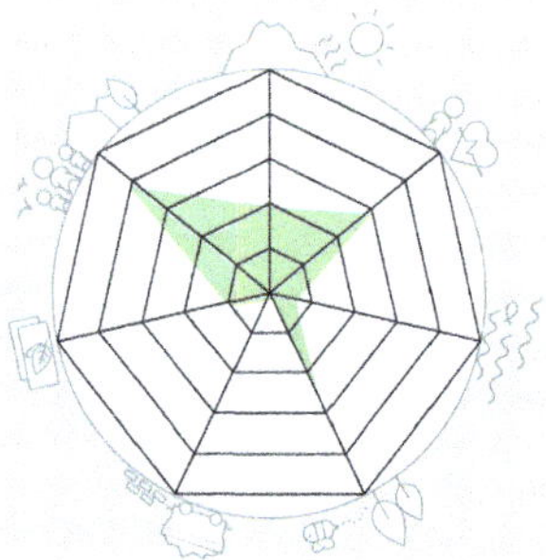

**Type**
Technological

**Building part**
Outdoor equipment

**Impact category**
Community

**Material**
Vegetation and/or water

**Intervention**
N.D.

**Scale**
Site

**Strategy**
Bioclimatic,
Well-being of workers

**Description**

Design organic open areas that are well-integrated into the built environment.

Open spaces should not be the direct consequence of the areas resulting from the design of the volumes. Rather, they must be a system of elements that feed each other as an essential element of the design process.

Within the area master plan, green spaces "define both the current and future performance of a factory's urban development concept" (Reichardt, 2015).

Therefore green spaces, pavements, and footpaths should be arranged per hierarchy and conceived within the general layout. This can be achieved ensuring the built and void lines continuity, or the choice of materials to use in the paths, furniture and vegetations.

**Practices**

a) open space re-design in front of the offices
A25 - Felsineo Veg (IT, 2017)

b) tiny square for employees and events
A1 - Melinda (IT, 2014); B2 - Cantina Pizzolato (IT, 2016)

c) paving patterns connected with the surrounding lines
A27 - Tangshan Organic (CH, 2016)

d) open spaces integrated with building layout
A18 - Iroha Village Factory (JP, 2018)

e) courtyards visually connected with natural landscape
A26 - Fratelli Beretta (U.S., 2016)

f) a courtyard connects the renovated and new buildings
B15 - CE.VI.V. (IT, 2018)

# SS 4.1 RAIN WATER COLLECTION

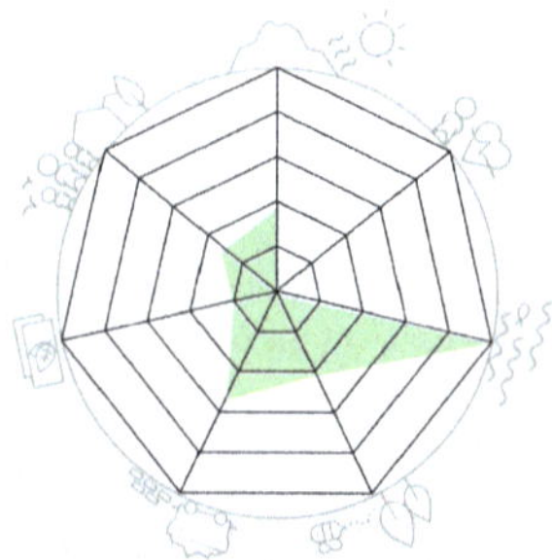

**Type**
Technological

**Building part**
Outdoor equipment

Top horizontal envelope

**Impact category**
Water resource

**Material**
Vegetation and/or water

**Intervention**
N.D.

**Scale**
Site

**Strategy**
Water saving

**Description**

Collecting and reusing rainwater can be beneficial in the industrial sector, where the large absorbing surfaces of roofs allow for the collection of large volumes of water.

This can be used for irrigation of open spaces, washing of loading and unloading areas, and even the operation of production machines in some cases. As a result, it is critical to identify and maximize all feasible catching surfaces, as well as connect them to appropriate collection systems.

The water can be stored in above-ground silos or underground tanks, and purified using naturalistic engineering techniques (e.g. phyto-depuration) or hydraulic groundwork (sedimentation of suspended particulates and nutrient removal) and/or rainwater infiltration (Comune di Bomporto, 2016).

Water treatment and collection must be carefully planned, to achieve positive landscape effects too (O.P.U.Q.P., 2015).

**Practices**

a) underground water tanks
A2 - Bio Pastificio IRIS (IT, 2015)
B5 - Macallan distillery (UK, 2018)

b) rain water collection as reservoir in case of fire
A9 - Celle ipogee Melinda (IT, 2004)

c) water basin in the site geometrical centre
A19 - Opificio Salpi (IT, 2010)

d) maximization of water catching surfaces
A16 - Mars Petcare (AU, 2015)
B1 - UC Davis Wine Brewery (CA, 2013)
B11 - Valprè (ZA, 2011)

e) rain water collection by concave roofing
A23 - Carnitalia (IT, 2016)

# SS 4.2 PERVIOUS SURFACES

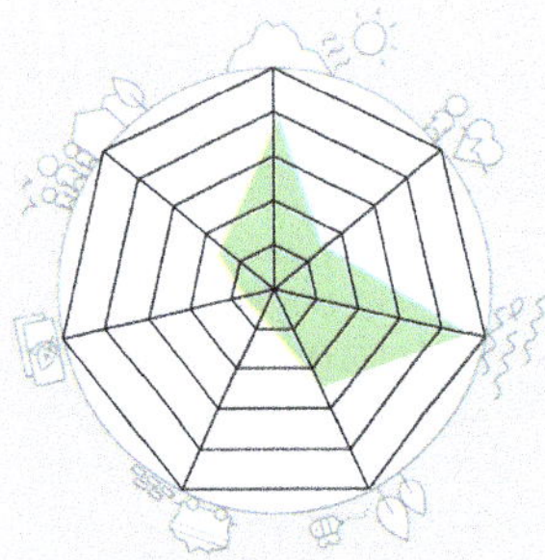

**Type**
Technological

**Building part**
Outdoor equipment

**Impact category**
Water resource

**Material**
Vegetation and/or water

**Intervention**
New Construction

**Scale**
Site

**Strategy**
Soil saving

**Description**

Provide good rainwater management, allowing it to infiltrate the soil and reach groundwater, or channel it to humid areas with suitable plants to absorb pollutants.

It is indeed critical to make the water drain through the soil rather than through evacuation pipes and channels (Grant, 1997). For this reason, whenever possible, use permeable surfaces for storage and parking spaces, while reducing hardscapes (O.P.U.Q.P., 2015; U.S. GBC, 2014).

Should the existing pavement be impervious, "de-sealing" or "de-paving" procedures can be performed to increase the percentage of permeable soil: e.g. replacing asphalt road pavement with self-locking blocks made of vibro-compressed concrete (Comune di Bomporto, 2016).

Further details can be found in the guide by Dessì et al. (2018), where common pavement surfaces are described along with the proprieties that most affect the urban micro-climate.

**Practices**

a) large green surfaces in the site
B1 - UC Davis Wine Brewery (CA, 2013)

b) open space design to maximize soil permeability
B11 - Valprè (ZA, 2011)

# SS 5.1 GREEN ROOF

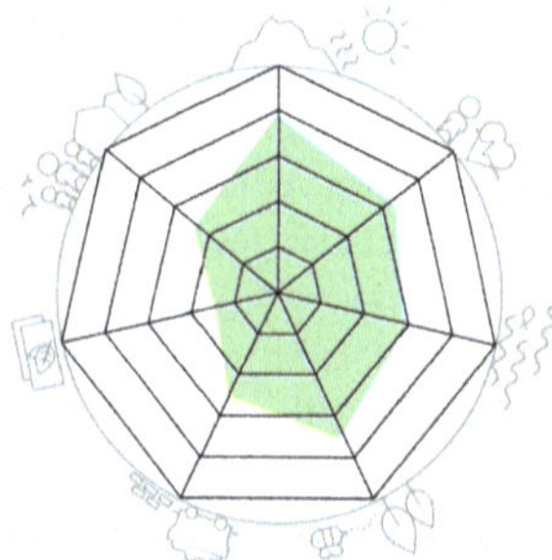

**Type**
Technological

**Building part**
Top horizontal envelope

**Impact category**
Climate change

**Material**
Vegetation and/or water

**Intervention**
New Construction

**Scale**
Building

**Strategy**
Bioclimatic

Visual mitigation

**Description**

The rooftop of industrial buildings is often suitable to be covered by vegetation because of its extension and geometry (Provincia di Bologna, 2009).
Green roofs, whether walkable or not, can help to improve the local micro climate by lessening the heat island effect.
Furthermore, they usually provide to the roof good inertia and thermal insulation, resulting in improved internal comfort and serving as a useful means for landscape integration.
Interestingly, a green roof can promote the preservation of biodiversity in urban areas as well as the absorption of polluting particles (Dessì et al., 2018).
On the other hand, the green roof limits the possibility of lighting from the top and the installation of photovoltaic and solar thermal systems. It also costs more than traditional industrial roofing, requires more maintenance and is less rapid to install.

**Practices**

a) non walkable green roof
A1 - Melinda (IT, 2014)
A2 - Bio Pastificio IRIS (IT, 2015)
A10 - Pellini Caffè (IT, 2006)
A18 - Iroha Village Factory (J, 2018)
A19 - Opificio Salpi (IT, 2010)

b) shrubs and trees on the roof
B3 - Cantina Antinori (IT, 2013)

# SS 5.2 GREEN FOLDED SURFACE

**Type**
Technological

**Building part**
Outdoor equipment

**Impact category**
Climate change

**Material**
Vegetation and/or water

**Intervention**
New Construction

**Scale**
Site

**Strategy**
Bioclimatic
Well-being of workers
Visual mitigation

**Description**

Earth or artificial hills covered by vegetation allow heat and water to be stored and slowly released, in most situations benefiting the summer micro climate. At the same time, they can produce a pleasant view of the factory in harmony with the natural surrounding or skyline.

In this regard, Dessì et al. (2018) present some beneficial water-control measures, ranging from floodable basins to "rain gardens". All of these help to improve the overall image of the site, while managing water and improving micro-climate.

Floodable basins, for example, can be normally exploited as parks for the use of workers and the population, with raised walkways for pedestrians, bicycles, skates, and be converted into rolling basins in the event of heavy rains.

**Practices**

a) green hill

A19 - Opificio Salpi (IT, 2010)

B5 - Macallan distillery (UK, 2018)

# SS 6.1 LIGHT POLLUTION PREVENTION

**Type**
Technological

**Building part**
Systems

**Impact category**
Community

**Material**
N.A.

**Intervention**
N.D.

**Scale**
Site

**Strategy**
Visual comfort

**Description**

Limit light pollution by adopting external luminaries with an adequate angle of light and, if possible, provide **programmed automatic switch-off systems. This helps to** reduce the negative impact on local fauna and facilitate the view of the starry sky.

**Furthermore, differentiating the external lighting systems** by function helps with improved visual comfort and energy saving. To this end, for instance, popular routes can be served by low-rise, turret systems (O.P.U.Q.P., 2015) which ensure better and more targeted enlightenment.

U.S. GBC (2014) particularly promotes street LED lighting that reduces the energy demands from neighbourhood infrastructure.

**Practices**

N.D.

# WE 1.1 OUTDOOR REUSE OF RAINWATER

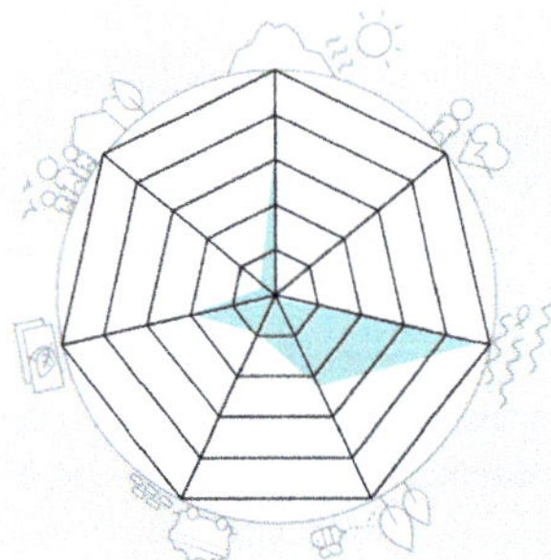

**Type**
Technological

**Building part**
Outdoor equipment

**Impact category**
Water resources

**Material**
Vegetation and/or water

**Intervention**
N.D.

**Scale**
Site

**Strategy**
Water saving

**Description**

**In green buildings, it is preferred efficiency of systems** before the reuse. The same applies to water systems.
**In landscape irrigation, for example, it means first designing water-wise landscaping and water-efficient** irrigation technology, then installing sub-meters to track and log irrigation trends, and only after using non-potable water (U.S. GBC, 2014).
In this regard, recycled rainwater that has been collected in ponds or cisterns (above or beneath the ground) can be **reused outside after purification according to the purpose.**
On the one hand, it can be functional to the management and maintenance of plant systems (irrigation of open green spaces, green roof or green façade), on the other hand, it can also be used in leisure spaces (water games to complement green space).
In any case, the goal is to consume as little drinking water as possible.

**Practices**

a) open space irrigation
A2 - Bio Pastificio IRIS (IT, 2015)
A16 - Mars Petcare (AU, 2015)
A19 - Opificio Salpi (IT, 2010)
B1 - UC Davis Wine Brewery (CA, 2013)

b) green roof irrigation
B5 - Macallan distillery (UK, 2018)

c) drop by drop irrigation for hedges
B2 - Cantina Pizzolato (IT, 2016)

# WE 2.1 WATER SAVING TECHNOLOGIES

**Type**
Technological

**Building part**
Systems

**Impact category**
Water resources

**Material**
N.A.

**Intervention**
N.D.

**Scale**
Building

Process

**Strategy**
Water saving

Green marketing

**Description**

Nowadays several water-saving technologies are available to reduce water consumption in the industry.

Some of them are based on reuse: for example, allowing the purification of production wastewater to be reused in the same or other manufacturing cycles.

Some others follow the efficiency principle and are implemented in innovative types of machinery, for example by using less water than required by traditional devices to perform the same work (Aster, 2014).

Also installing efficient plumbing fixtures can help, including high-efficiency lavatories, kitchen sinks, and showers, dual-flush toilets, waterless urinals, and composting toilets (U.S. GBC, 2014). When the installation of a new fixture is too costly, at least new flush valves or flow restrictions can be used.

When the local standards allow it, non-potable water as captured rainwater, treated grey water, or municipally-provided reclaimed water can be used for flush fixtures and saving potable water.

**Practices**

N.D.

# WE 4.1 WATER MANAGEMENT SYSTEMS

**Type**
Technological

**Building part**
Systems

**Impact category**
Water resources

**Material**
N.A.

**Intervention**
N.D.

**Scale**
Building
Process

**Strategy**
Water saving

**Description**

The first step to improve the performance of the production system and save resources is to identify inefficiencies and waste. Constant monitoring and effective management systems can help to this end.

It is recommended to install sub-meters to measure indoor water usage (i.e., monitor the data to track consumption trends, determine fixture performance, and pinpoint leaks) and also to raise employees' awareness about water.

Not only it is critical to keep track of civil water use, but also to track processes and machinery to spot potential plant enhancements, and resolve leaks or network obsolescence. Monitoring is also necessary for the development of ad hoc withdrawal strategies that minimize the impact on local water bodies.

Once the status of the water system is known and connected with the overall technical system, water-saving strategies can be more effectively implemented. For example: by installing high-performance irrigation systems, such as drip systems and bubbler distribution systems, channel water directly to root systems, and weather-based irrigation controllers (U.S. GBC, 2014).

**Practices**

a) water sub-meters system
A11 - Amadori (IT, 2016)

b) multiple withdrawal points to mitigate the interferences with the river
B6 - Dalmunach distillery (UK, 2015)

c) "pulsed output water meters" linked with a leakage detection system
B13 - St. James Gate Brewery (UK, 2011)

# EA 2.1 VENTILATED FAÇADE

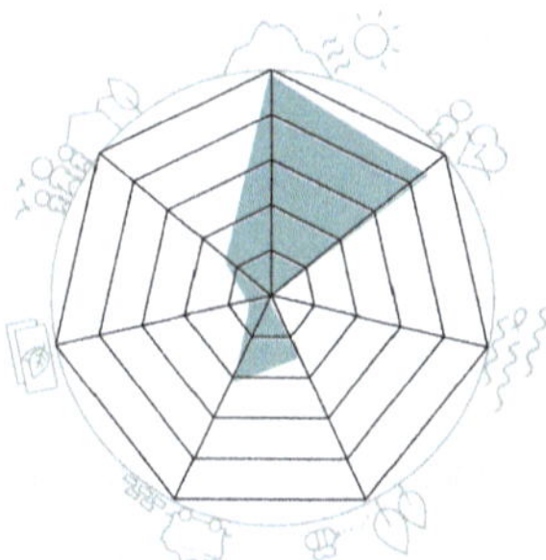

**Type**
Technological

**Building part**
Vertical envelope

**Impact category**
Climate change

**Material**
N.D.

**Intervention**
N.D.

**Scale**
Building

**Strategy**
Energy efficiency
Well-being of workers
Visual mitigation

**Description**

A ventilated façade is a perimeter wall system with an air gap between the load-bearing framework (including any functional insulating layer) and the cladding (often installed on a light metal structure). The system is usually beneficial to both the energy performance of the envelope and the visual quality of the building.

It is fast to install thus it is especially advantageous in retrofitting existing buildings.

Concerning the thermal benefits, the cavity allows the passage of air determined by a natural chimney effect, and in summer this shields the underlying structure from thermal changes by extracting excess heat due to solar radiation. Its contribution to the overall energy performance of the envelope largely depends on the type of cavity (ventilated or poorly ventilated).

Pay particular attention to any undesired effects, such as acoustic and fire issues that may be complicated by the cavity.

Finally, due to the variety of material and finishing options available on the market, this technology is a good choice for the restyling of façades from an aesthetic perspective.

**Practices**

a) perforated, mesh or expanded metal sheets
A1 - Melinda (IT, 2014)
A3 - Almazara Olisur (CL, 2009)

b) wooden panels or boards
A26 - Fratelli Beretta (U.S., 2016)
B9 - Cantina di Collemassari (IT, 2005)

# EA 2.2 SHADING SYSTEMS

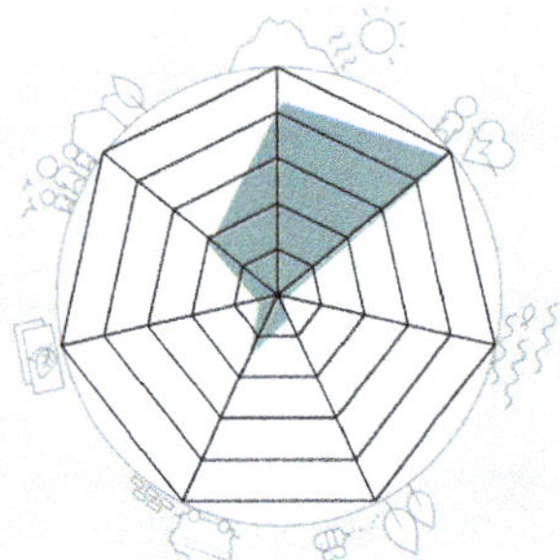

**Type**
Technological

**Building part**
External partition

**Impact category**
Health and Well-being

**Material**
N.D.

**Intervention**
N.D.

**Scale**
Building

**Strategy**
Energy efficiency

Visual comfort

Brand image

**Description**

Shading systems (e.g., discontinuous screens, porches and projections, or trees) serve to shield the solar radiation in the summer (Grant, 1997) and avoid unwanted heat gain by the passive behaviour of the building. U.S. GBC (2014) stresses considering the context to adopt suitable shading: for example based on adjacent buildings that might shade and insulate each other.

Given that in factories the transparent part is mainly in the roof, pay attention to avoid unshielded zenith openings, but at the same time ensure adequate levels of daylight (Province of Bologna, 2008).

Shading elements are also helpful to qualify the façades of buildings: sunshades, porches and overhangs, and micro-perforated sheets are among the most common systems.

**Practices**

a) perforated metal sheet to filter light
A1 - Melinda (IT, 2014)

b) horizontal brise-soleil A10 - Pellini Caffè (IT, 2006); A15 - Brown Sugar Factory (CH, 2016); B7 - Carlsberg (FR, 2014); B9 - Cantina di Collemassari (IT, 2005)

c) photovoltaic as horizontal brise-soleil
A13 - C.O.C.E.A. (IT, 2015)

d) shading by sticks bundles to dry and then used in the process A18 - Iroha Village Factory (J, 2018)

e) vertical timbers to filter both light and externals sight
A26 - Fratelli Beretta (U.S., 2016); A28 - Mulino Agri '90 (IT, 2011)

f) porch on east and west sides
B1 - UC Davis Wine Brewery (CA, 2013)

g) vertical brise-soleil and overhanging roof
B12 - Bavaria Brewery (CO, 2011); B13 - St. James Gate Brewery (UK, 2011)

# EA 2.3 THERMAL INSULATION

**Type**
Technological

**Building part**
Top horizontal envelope
Low horizontal envelope

**Impact category**
Climate change

**Material**
N.D.

**Intervention**
N.D.

**Scale**
Building

**Strategy**
Energy efficiency
Well-being of workers
Green marketing

**Description**

Providing a proper thermal insulation system is now common in new or recently renovated buildings due to regulations and financial incentives, as well as the ease of implementation and relatively low costs.

In factories, where windows are limited, the tactic refers to opaque envelope where there is large use of prefabs layered with insulating panels. Especially with these fast-assembly systems, it is critical paying attention to joints, to limit thermal bridges (Provincia di Bologna, 2009) and to adequately cope transmittance parameters with thermal inertia for summer comfort.

Furthermore, adequate envelope airtightness is required to ensure that the building is suitably weatherized. This is especially true in industries, where doors must be thermally safe due to their frequent opening: a series of fast-closing doors can greatly contribute (Duflou et al., 2012).

**Practices**

a)thermal transmittance half than standard
A2 - Bio Pastificio IRIS (IT, 2015)

b) cold area insulated by panels in PIR plus rock-wool
A23 - Carnitalia (IT, 2016)

c) prefabricated concrete panels with inner insulation
A24 - Bell & Evans (U.S., ongoing)

d) differentiated insulation between factory and offices
A25 - Felsineo Veg (IT, 2017)

e) prefab insulating panels embedding cold/hot water pipes
B1 - UC Davis Wine Brewery (CA, 2013)

f) double façade with corridor creating a thermal buffer
B12 - Bavaria Brewery (CO, 2011)

g) excellent envelope air tightness, door included
B13 - St. James Gate Brewery (UK, 2011)

# EA 2.4 THERMAL MASS

**Type**
Technological

**Building part**
Vertical envelope
Top horizontal envelope
Low horizontal envelope

**Impact category**
Climate change

**Material**
Wood
Concrete

**Intervention**
N.D.

**Scale**
Building

**Strategy**
Energy efficiency
Well-being of workers

**Description**

In addition to the thermal transmittance performance of the construction packages, consider the phase shift factor of the thermal wave. This is the time difference between the time when the maximum temperature is recorded on the external surface of the structure and the time when the maximum temperature is recorded on the internal surface of the same.

Massive construction methods improve the building's thermal inertia, which can be used to keep rooms cold or, on the other hand, to store heat through direct radiation and benefit from its progressive release later.

This performance can be combined with thermal insulation in retrofitting: in general, the higher the insulation density, the greater the phase shift.

The green roof is a good example in hot locations since it reduces the quantity of solar radiation that enters the building and so enhances the internal thermo-hygrometric comfort (Chowdhury et al., 2017).

**Practices**

a) green roof
A1 - Melinda (IT, 2014)
A19 - Opificio Salpi (IT, 2010)
B3 - Cantina Antinori (IT, 2013)
B5 - Macallan distillery (UK, 2018)

b) massive prefab concrete walls
A24 - Bell & Evans (U.S., ongoing)
B1 - UC Davis Wine Brewery (CA, 2013)

# EA 2.5 GROUND THERMAL PROPRIETIES

**Type**
Technological

**Building part**
Top horizontal envelope
Low horizontal envelope

**Impact category**
Climate change

**Material**
Vegetation and/or water

**Intervention**
N.D.

**Scale**
Building

**Strategy**
Bioclimatic
Visual mitigation

**Description**

Underground or semi-underground industrial buildings can have several benefits:

i) environmental, such as the capacity to regulate temperature and humidity regardless external conditions due to the thermal mass of the ground;

ii) and aesthetic-perceptive, such as the ability to partially hide the complex with respect to the scenery.

The strategy is particularly useful for agri-food industrial buildings, which frequently require a great amount of energy to store products in cool cells and are often set in high-quality landscape surroundings or close to rural areas. For these reasons, many agri-food companies have rediscovered the ancient technique of underground building for food preservation in their new facility (Mazarrón, Cid-Falceto, Cañas, 2012).

**Practices**

a) underground cooling cells for food preservation
A9 - Celle ipogee Melinda (IT, 2004)
A13 - C.O.CE.A. (IT, 2015)

b) semi-underground buildings
A18 - Iroha Village Factory (J, 2018); A19 - Opificio Salpi (IT, 2010); B3 - Cantina Antinori (IT, 2013); B5 - Macallan distillery (UK, 2018); B9 - Cantina di Collemassari (IT, 2005); B10 - Cantina Pieve Vecchia (IT, 2012)

c) underground parts for storage, exterior parts for $CO_2$ emitting processes
B4 - Bodegas Portia (ES, 2010)

d) underground building
A9 - Celle ipogee Melinda (IT, 2004)

# EA 2.6 GREEN FAÇADE

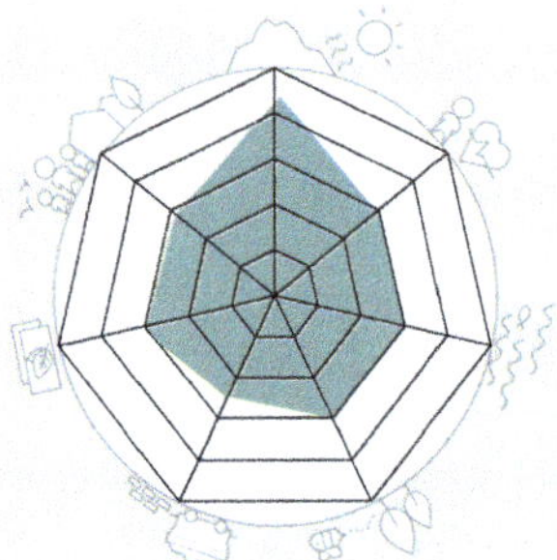

**Type**
Technological

**Building part**
Vertical envelope

**Impact category**
Climate change

**Material**
Vegetation and/or water

**Intervention**
N.D.

**Scale**
Building

**Strategy**
Bioclimatic

Well-being of workers

Visual mitigation

**Description**

A green façade for a building or portions of it can increase not only its aesthetic impact, but also its energy efficiency. The benefits of the green wall on the building's energy performance are comparable to those of the ventilated wall, likewise its functions as a "second skin".

The green wall prevents direct radiation from the wall, which does not heat up and does not radiate heat to the 'interior. Also, the thermal mass of the wall increases, increasing the heat wave shift. In addition, since it captures fine dust through the foliar apparatus the local air quality can result improved too (Dessì et al., 2018).

However, costs and needs for maintenance are usually higher than traditional mineral claddings, even if it largely varies from the type of green wall (extensive or intensive). In rural areas, the strategy is frequently used to camouflage buildings that are not in harmony with their surroundings, whereas in metropolitan areas, it is typically applied to convey environmental sustainability messages to the public (i.e. mostly with aesthetic and ornamental purposes).

**Practices**

N.D.

# EA 2.7 DEVICES FOR EFFICIENT HEATING

**Type**
Technological

**Building part**
Systems

**Impact category**
Climate change

**Material**
N.A.

**Intervention**
N.D.

**Scale**
Building

**Strategy**
Energy Efficiency

Well-being of workers

**Description**

Both in new construction and retrofitting, several tools and technologies are nowadays available to reduce energy demand from heating systems and in turn to limit negative emissions in the atmosphere.

One easy-to-apply and widely spread example is the air destratifier: that is, an air recirculation system capable of reducing the stratification effect of heat typical of production environments that are large and very high (Delpiano and Tombetti, 2011).

In a nutshell, these are fans that re-direct heat to the floor, preventing heat flow from the ceiling to the outside and lowering heat loss. As a result of the improved overall efficiency of the heating system, the building heats up faster and conserves energy in the process of generating hot air.

**Practices**

N.D.

# EA 3.1 TIMED OR MEN-PRESENCE ACTIVATION SYSTEMS

**Type**
Technological

**Building part**
Systems

**Impact category**
Climate change

**Material**
N.A.

**Intervention**
N.D.

**Scale**
Building

**Strategy**
Automatic system

Visual comfort

**Description**

Install smart or automatically timed systems, or systems that activate when they sense human presence, to save energy for lighting and heating, especially in automated warehouses and storage facilities where personnel enter seldom and/or objects do not require constant climatic conditions.

Mueller et al. (2014) recommend installing presence sensors (e.g. Infrared sensors) or monitoring thermo-hygrometric conditions to regulate the heating/lighting systems rather than over-sizing the plant systems.

**Practices**

a) Infra-Red sensors activating lights

A7 - Kerrygold (UK, 2008)

B11 - Valprè (ZA, 2011)

b) timed lighting to reduce waste and not interfere with fermentation process

B2 - Cantina Pizzolato (IT, 2016)

# EA 3.2 ADVANCED ENERGY MANAGEMENT

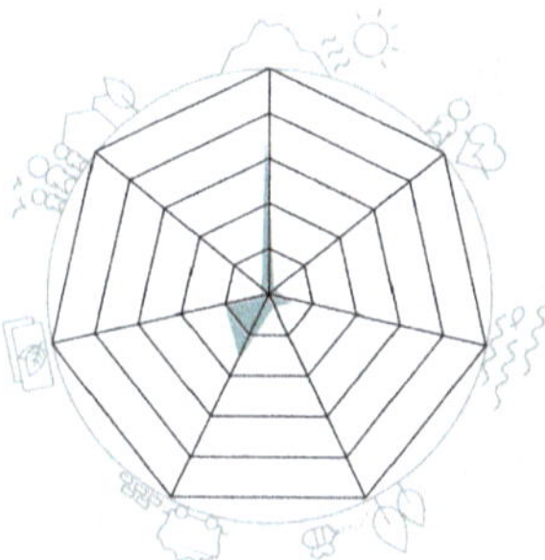

**Type**
Technological

**Building part**
Systems

**Impact category**
Climate change

**Material**
N.A.

**Intervention**
N.D.

**Scale**
Building

**Strategy**
Energy Efficiency

**Description**

Install load monitoring systems to keep track of how much electricity, gas, compressed air, technical heating, and water are used, so you can plan ahead and perhaps balance demand.

For example, users/machinery with lower priority (excess/double machinery or high thermal inertia furnaces) may be temporarily turned off to stay within the fixed quota if the workload exceeds a particular level every 15 minutes (Duflou et al., 2012).

The tactic also relates to Demand Response (DR) programmes, which encourage electricity customers to reduce their usage during peak demand times. DR helps utilities to optimize their supply-side energy generation and delivery systems: for example, an incentive program can reward consumers who agree to change their usage patterns when the utility company sends an alert (to the building operator or the building automation system) announcing a curtailment event (whose maximum number in specific time frames are set) (U.S. GBC, 2014).

In this framework, energy re-phasing is another interesting practice: it consists in limiting the withdrawals of inductive reactive energy by power factor correction, and supplying the system with capacitive reactive energy by means of batteries of capacitors that allow not to demand energy from public networks (Delpiano e Trombetti, 2011).

**Practices**

N.D.

# EA 4.1 LOCAL ENERGY PROVIDER

**Type**
Technological

**Building part**
Systems

**Impact category**
Climate change

**Material**
N.A.

**Intervention**
N.D.

**Scale**
Building

**Strategy**
Energy efficiency
Community engagement
Green marketing

**Description**

Enhance the presence of a plant in the area from an energy perspective. In fact, the complex can operate as a local energy supplier or, on the other hand, as a buyer of renewable energy supplied by local residents.

As a result, the factories would become a decentralised unit of the energy distribution network, having a positive impact on the surrounding, even in the event of a power outage (i.e. blackout) (Herrmann et al., 2015).

An example of a factory producing a surplus of renewable energy and heat is the J. Schmalz, in Glatten (Germany).

**Practices**

N.D.

# EA 4.2 ON-SITE ENERGY PRODUCTION

**Description**

On-site energy production can avoid utilities to build additional power generation facilities, transmission lines, and distribution stations. As a result, some negative environmental effects of energy infrastructure and consumption can be limited (U.S. GBC, 2014).

In order to evaluate the convenience of this tactic, Mueller et al. (2014) suggest considering the type of energy that the company requires, including the transformation on the site; the existing supply infrastructures and also the energy conservation/storage requirements.

**Type**
Technological

**Building part**
Systems

**Impact category**
Climate change

**Material**
N.A.

**Intervention**
N.D.

**Scale**
Building

**Strategy**
Energy Efficiency

**Practices**

N.D.

# EA 4.3 SCALABLE TECHNICAL BUILDING SYSTEMS

**Type**
Technological

**Building part**
Systems

**Impact category**
Climate change

**Material**
N.A.

**Intervention**
New construction

**Scale**
Building

**Strategy**
Energy efficiency

**Description**

Design scalable and modular Technical Building Systems (TBSs) as suggested by sustainability and resilience principles.

Regarding in particular the energy supply networks, it is a question of conceiving TBSs as open systems in which new elements and devices can be easily added or removed. For instance, by plugging or unplugging devices, according to the fluctuation of demand (reduction or expansion of the same throughout the day or over a different period) (Hermann et al., 2015). Furthermore, modular TBSs can act as network balancing components.

If the additional elements are renewable energy sources, such as solar, wind, hydroelectric, or geothermal, storage technologies like as batteries or buffers become critical parts of the system: Solvis GmbH, Braunschweig (D), is a great example in this field.

**Practices**

N.D.

# EA 5.1 SOLAR ENERGY

**Type**
Technological

**Building part**
Systems

**Impact category**
Climate change

**Material**
N.A.

**Intervention**
N.D.

**Scale**
Building

**Strategy**
Renewable resources
Green marketing

**Description**

Integrate solar energy into industrial processes to lower the economic and environmental costs of fossil fuels, hence **boosting the company's competitiveness. In this field the system flexibility and scalability is crucial.**

Photovoltaic (PV) must be preferably installed in already urbanised areas, unused lot or **any surface that is difficult to** dedicate to other uses, in order to prevent the consumption of agricultural and natural areas. In factories, it is suggested placing it on building roofs or canopies/shelters for parking (O.P.U.Q.P., 2015).

In addition to photovoltaic that is now widespread at industrial level, also solar panels technology has spread in the sector (e.g. high-vacuum collectors) (Koçak et al., 2020) and several storage technologies are now available, from underground thermal energy storage to packed beds. The agri-food industry, in particular, can take advantage of solar energy for those activities ranging from 30 to 120 °C (e.g. drying; washing; pasteurization).

**Practices**

a) rooftop photovoltaic
A2 - Bio Pastificio IRIS (IT, 2015); A7 - Kerrygold (UK, 2008); A11 - Amadori (IT, 2016); A28 - Mulino Agri '90 (IT, 2011); B1 - UC Davis Wine Brewery (CA, 2013); B2 - Cantina Pizzolato (IT, 2016); B17 - Distilleria Zanin (IT, 2013)

b) photovoltaic on façade
A5 - Grandi Salumifici Italiani (IT, 2013); A13 - C.O.CE.A. (IT, 2015)

c) photovoltaic on the ground
B11 - Valprè (ZA, 2011)

d) energy self-sufficiency
A6 - Solare Manufaktur Peter (DE, 2012)

e) combination of diverse photovoltaic types
A17 - Frito-Lay (TX, 2011)

f) thermal solar panels
A19 - Opificio Salpi (IT, 2010); B11 - Valprè (ZA, 2011)

# EA 5.2 GEOTHERMAL ENERGY

**Type**
Technological

**Building part**
Systems

**Impact category**
Climate change

**Material**
N.A.

**Intervention**
New construction

**Scale**
Building

**Strategy**
Renewable resources
Green marketing

**Description**

Select the most suitable renewable energy sources for the production site, based on the local climate (Provincia di Bologna, 2008) and the specific needs of the company.
Heat exchangers and geothermal systems, which draw heat from the subsoil or the water, can be especially useful in some industrial activities (Grant, 1997).
Although the large footprint of industrial buildings suggests the use of horizontal probe systems, it will be professionals (including thermal technician and geologist) who will indicate the most appropriate type on a case-by-case basis.

**Practices**

a) geothermal probes
A3 - Almazara Olisur (CL, 2009)

b) water filtered through rock at 100 meters underground, used to cooling down compressors
A9 - Celle ipogee Melinda (IT, 2004)

c) predisposition for geothermal probes
A1 - Melinda (IT, 2014)

# EA 5.3 ENERGY FROM BIOMASS

**Type**
Technological

**Building part**
Systems

**Impact category**
Material resources

**Material**
N.A.

**Intervention**
N.D.

**Scale**
Building

**Strategy**
Renewable resources

Green marketing

**Description**

Integrate renewable energy sources that exploit local climatic resources with biomass plants, which use the by-products of internal processing in the area (Provincia di Bologna, 2008) or those derived from circular economy chains.

The agri-food industry is particularly interested in this tactic, as demonstrated by several virtuous processes of close industrial cycles and enhancement of by-products (e.g., kernels or skins).

The Emilia-Romagna region has even launched food-related research aimed at developing procedures for the transformation or conversion of animal and vegetable by-products into biofuels (Osservatorio GreenER, 2014).

**Practices**

a) biomass to steam generation
A17 - Frito-Lay (TX, 2011)

b) biomass for cogeneration
B5 - Macallan distillery (UK, 2018)

c) biomass from company by-products to heat
B14 - Bombay Sapphire Distillery (UK, 2013)

# EA 5.4 COGENERATION

**Type**
Technological

**Building part**
Systems

**Impact category**
Green economy

**Material**
N.A.

**Intervention**
N.D.

**Scale**
Building

**Strategy**
Energy efficiency

Green marketing

**Description**

Cogeneration is the combined and simultaneous production of electrical and thermal energy from a single source of energy: typically, only 40% of the energy released by combustion in engines is converted into electricity, with the rest being wasted as heat. Cogeneration recovers this share, **boosting the plant's efficiency.**

When two sources of energy are required at the same time, cogeneration comes useful. When compared to traditional natural gas, the type of fuel utilized can improve the plant's "sustainability": biogas, biomass, solar energy, or even hydrogen. It can be used by a single complex (micro-cogeneration) or by centralized systems that supply users in the surrounding area.

If associated with a refrigeration machine, it is called Trigeneration: a plant that produces electricity and thermal energy, both in the form of heat and chilled water, useful for air conditioning or for industrial processes that require low temperatures.

Philipp et al. (2016) and the Food Industry Sustainability Strategy of United Kingdom (DEFRA, 2006) both suggest cogeneration in the agri-food industry.

**Practices**

a) cogeneration plant

A7 - Kerrygold (UK, 2008)

B5 - Macallan distillery (UK, 2018)

b) cogeneration plant for both the company and neighbours

B13 - St. James Gate Brewery (UK, 2011)

# EA 5.5 WIND ENERGY

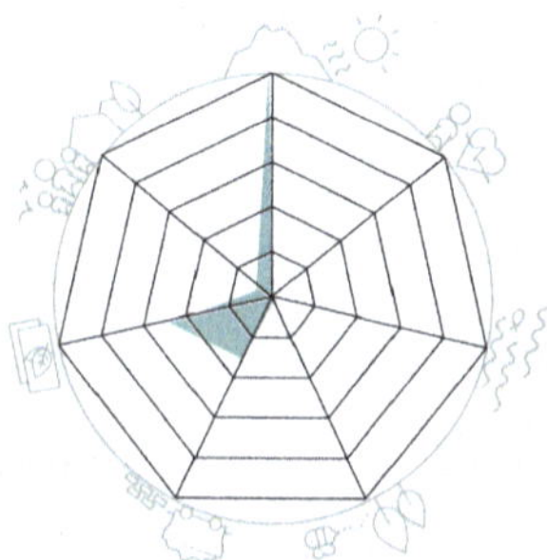

**Description**

Investigate the possibility of using wind energy, based on the climatic conditions of the site (Provincia di Bologna, 2008) and the specific needs of the company.

Both off-site wind farms (in some circumstances even off-shore) or on-site wind farms (usually a small wind farm) can provide this type of clean energy (Delpiano and Trombetti 2011).

**Type**
Technological

**Building part**
Systems

**Impact category**
Climate change

**Material**
N.A.

**Intervention**
N.D.

**Scale**
Site

**Strategy**
Renewable resources

**Practices**
N.D.

# EA 5.6 OFF-SITE GREEN ENERGY PRODUCTION / PURCHASE

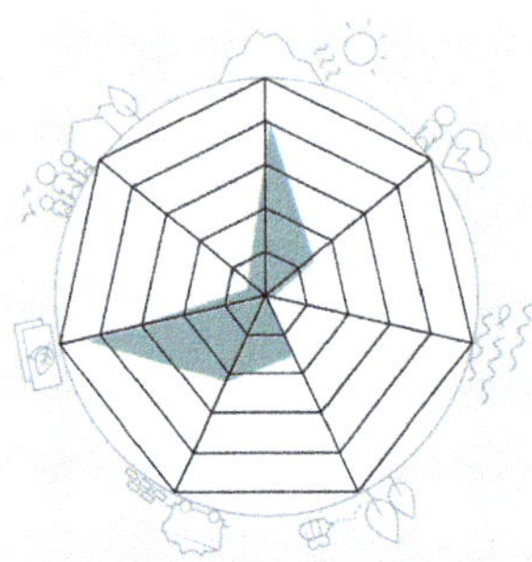

**Type**
Management

**Building part**
Systems

**Impact category**
Green economy

**Material**
N.A.

**Intervention**
N.D.

**Scale**
N.A.

**Strategy**
Renewable resources
Green marketing

**Description**

The tactic helps to reduce the ecological footprint of the company by green energy, as supported by the EC COM(2016) 860 - Clean Energy for All Europeans.

In addition to off-site or on-site energy supply from renewable sources, a corporation can compensate for its environmental negative impact by purchasing green certificates or carbon credits.

Green certificates, for example, correspond to a certain amount of $CO_2$ emissions: if a plant produces energy by emitting less $CO_2$ than a plant fuelled with fossil fuels (oil, natural gas, coal) would have done because "from renewable sources," the manager receives green certificates, which he can resell (at market prices) to industries or activities that are required to produce a certain amount of energy.

Instead, off-site renewable energy is typically purchased at a premium price per kilowatt-hour from a utility or a provider of Renewable Energy Certificates (RECs) (U.S. GBC, 2014).

**Practices**

a) energy from off-site photovoltaic
A19 - Opificio Salpi (IT, 2010)
B14 - Bombay Sapphire Distillery (UK, 2013)

b) energy from biogas
B7 - Carlsberg (FR, 2014)

c) hydroelectric power from 6kW turbine in the river
B14 - Bombay Sapphire Distillery (UK, 2013)

d) energy from off-site wind farm
A3 - Almazara Olisur (CL, 2009)

# EA 6.1 GHG FREE REFRIGERATION

**Type**
Technological

**Building part**
Systems

**Impact category**
Climate change

**Material**
N.A.

**Intervention**
N.D.

**Scale**
Process

**Strategy**
Air pollution reduction

**Description**

Use environmentally friendly refrigerant fluids (Provincia di Bologna, 2008) to reduce the emission of climate-altering gases into the atmosphere (CFC and HCFC are among the most harmful). In the 1980s, research demonstrated that certain refrigerants for building systems were depleting stratospheric ozone, which is a gas that protects human health and the environment by absorbing harmful UV radiation. As a consequence, these refrigerants contribute to climate change. So, in the 1990s, the Montreal Protocol banned the production of chlorofluorocarbon (CFC) refrigerants and established the phasing out of HCFC refrigerants.

Today, the solar cooling machine is particularly sustainable, or also the use of refrigerant fluid from local district cooling systems.

Solar cooling consists in combining a refrigeration machine with solar thermal panels. Considering the costs and sizes of the refrigeration units (over 20 kiloWatt), solar cooling is especially convenient for centralized air conditioning or refrigeration systems, or for large areas where the cold air/ water is brought into the individual parts of the building through a duct system or a distribution network.

**Practices**

a) plant free from GHG
A6 - Solare Manufaktur Peter (DE, 2012)

b) cooling system with natural ammonia
A24 - Bell & Evans (U.S., ongoing)

c) solar cooling
B1 - UC Davis Wine Brewery (CA, 2013)

d) glycol from an existing district cooling system
B13 - St. James Gate Brewery (UK, 2011)

# MR 1.1 RECYCLABLE STRUCTURE

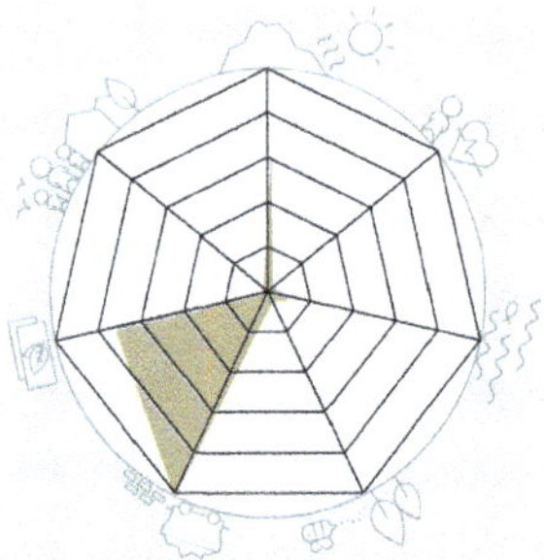

**Type**
Technological

**Building part**
Load-bearing structure

**Impact category**
Material resources

**Material**
Wood

Metals

**Intervention**
New construction

**Scale**
Building

**Strategy**
New material saving

**Description**

Designing from the beginning a structure that can be reused or recycled in the future helps to reduce the environmental impact of the building over its whole life cycle, as reuse is the best form of material conservation (U.S. GBC, 2014).
This reduces the need for new materials to be created and limits their processing environmental impact.
To this end, it is preferable to use dry construction systems, which ensure the possibility of dismantling the building at the end of its useful life and reusing its main components, but also involve faster construction times (Provincia di Bologna, 2009).
When using standard components, their likely reuse in other contexts is higher.

**Practices**

a) steel structure
A1 - Melinda (IT, 2014)
A15 - Brown Sugar Factory (CH, 2016)

b) mainly wooden-based dry construction techniques
A6 - Solare Manufaktur Peter (DE, 2012)
A23 - Carnitalia (IT, 2016)

c) standardised American technology made of steel structure and wooden-based dry walls
A26 - Fratelli Beretta (U.S., 2016)

# MR 1.2 MODULAR STRUCTURE AND/OR COMPONENTS

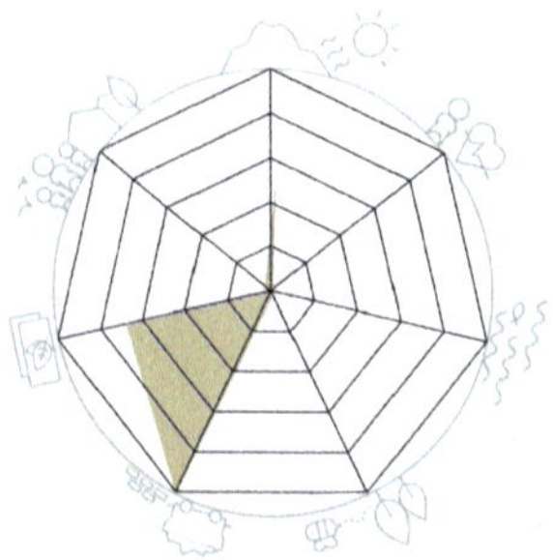

**Type**
Technological

**Building part**
Load-bearing structure
Internal partition
Vertical envelope

**Impact category**
Material resources

**Material**
N.D.

**Intervention**
New construction

**Scale**
Building

**Strategy**
New material saving
Flexibility

**Description**

Design buildings that can easily transform over time.
Since the spatial characteristics of manufacturing buildings depend on the specific needs of each company and the systems in use (Provincia di Bologna, 2009), the flexibility of the spaces adds value to a property by making it easier to adapt to the company's future needs or the needs of other companies that may take over.
Dry construction solutions can be used to design the building as a modular and expandable structure (Herrmann et al., 2015). Also, conceive the plan as a free space with large spans that reduce the spatial constraints and allow several configurations.

**Practices**

a) modular structure that can be enlarged in the future
B5 - Macallan distillery (UK, 2018)

b) large spans free from pillars that facilitate flexible use (free plan)
A17 - Frito-Lay (TX, 2011)
B17 - Distilleria Zanin (IT, 2013)

# MR 1.3 BUILDING RETROFITTING

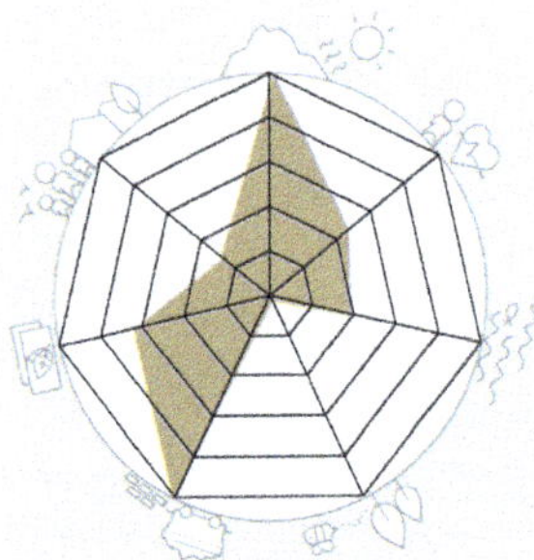

**Type**
Technological

**Building part**
N.D.

**Impact category**
Climate change

**Material**
N.D.

**Intervention**
Renovation

**Scale**
Building

**Strategy**
Energy Efficiency
Green marketing

**Description**

Improve the technical performance of an existing asset to **reduce building energy needs through energy retrofitting,** as well as to limit land consumption, the use of new natural resources, avoid demolition costs (both economic and environmental), to improve seismic safety, and even to enhance the building's aesthetic characteristics (O.P.U.Q.P., 2015).

The concept is now widely acknowledged and implemented in Europe, with great institutional support through tax breaks and incentives.

**Practices**

a) energy retrofitting
A5 - Grandi Salumifici Italiani (IT, 2013)

b) deep renovation of existing building
A22 - Caffè Milani (IT, 2017)
B14 - Bombay Sapphire Distillery (UK, 2013)
B17 - Distilleria Zanin (IT, 2013)

c) seismic and energy retrofitting
A23 - Carnitalia (IT, 2016)
A25 - Felsineo Veg (IT, 2017)

d) renovation of historic heritage
B16 - Roe & Co Distillery (IE, 2019)

# MR 1.4 DURABILITY AND LOW MAINTENANCE

**Type**
Technological

**Building part**
N.D.

**Impact category**
Material resources

**Material**
N.D.

**Intervention**
N.D.

**Scale**
Building

**Strategy**
New material saving

**Description**

Prefer materials that are long-lasting and low-maintenance. A variety of criteria influence the selection of materials and project components, including their aesthetic qualities, prices, and transportation circumstances.
When just looking at costs, however, keep in mind that the potential additional cost of resistant and durable materials must be weighed against the long-term benefits, which include the type and degree of maintenance necessary.

**Practices**

a) materials selected for their excellent durability
A24 - Bell & Evans (U.S., ongoing)

b) materials selected for their low maintenance needs
B12 - Bavaria Brewery (CO, 2011)

# MR 2.1 LOW LIFE CYCLE IMPACT

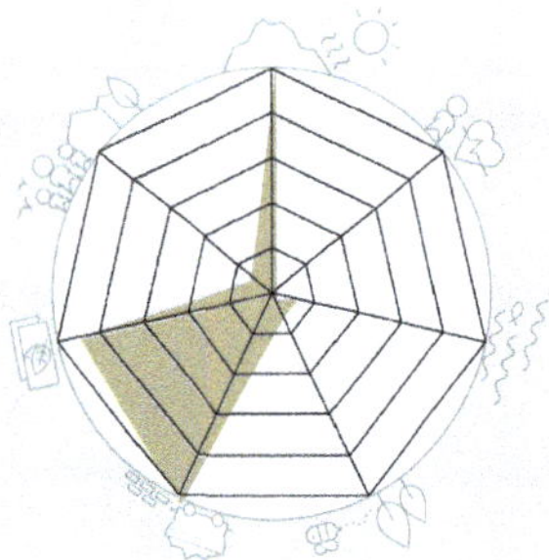

**Type**
Technological

**Building part**
N.D.

**Impact category**
Material resources

**Material**
N.D.

**Intervention**
N.D.

**Scale**
Building

**Strategy**
Energy efficiency,
Green marketing

**Description**

Reduce the facility's environmental impact throughout its life cycle by addressing not just the operational phase (energy efficiency, plant engineering), but also the factory's construction and decommissioning stages.

Recent studies on the environmental balance of buildings show that the weight of the operational phase is decreasing in comparison to a significant increase in the material production stages, whose environmental performance on site is offset by consistent amounts of energy required in the previous phases.

As a result, it is recommended to select materials with low energy and emissions incorporated throughout the manufacturing process, with a preference for those with lower embodied energy (i.e., materials using low energy consumption production processes) (Province of Bologna, 2008).

If available, Life Cycle Assessment (LCA) studies on products can represent valid assistance to this end.

**Practices**

a) wooden cladding that requires low maintenance
A19 - Opificio Salpi (IT, 2010)

b) over 95% of certified materials
A21 - Van Gelder (NL, 2019)

c) low impact materials and also anti-bacterial surfaces
A25 - Felsineo Veg (IT, 2017)

d) concrete masonry blocks with reduced share of cement and foundations with 50% less cement than usual
B1 - UC Davis Wine Brewery (CA, 2013)

# MR 2.2 RECYCLED/RECYCLABLE MATERIALS

**Type**
Technological

**Building part**
N.D.

**Impact category**
Material resources

**Material**
N.D.

**Intervention**
N.D.

**Scale**
Building

**Strategy**
New material saving
Green marketing

**Description**

Reduce the environmental impact of the building throughout its life cycle by using recyclable materials or recovered inert waste, recycling and reuse procedures for technical elements, and selective demolitions (Provincia di Bologna, 2008).

Recycling and reuse are essential for all construction materials, from the structure to the envelope (Herrmann et al., 2015).

Moreover, in some historical buildings, reusing a material entails not only lowering environmental costs, but also preserving its cultural value.

**Practices**

a) reuse of materials from the demolished building
B6 - Dalmunach distillery (UK, 2015)
B14 - Bombay Sapphire Distillery (UK, 2013)

# MR 3.1 LOCAL MATERIALS

**Type**
Technological

**Building part**
N.D.

**Impact category**
Green economy

**Material**
N.D.

**Intervention**
N.D.

**Scale**
Building

**Strategy**
Air pollution reduction
Community engagement
Green marketing

**Description**

Prefer materials produced close to the construction site, with low environmental transportation costs because of the distances from the building (Provincia di Bologna, 2008). Avoid products that require air freight as a general rule.
The use of local materials not only saves the economic and environmental expenses associated with product shipping, but also helps to support the local economy and, in some cases, improves visual coherence of the building by better integrating the product with its surroundings.
For example, use wooden brise-soleil from surrounding forests, or bricks that are typical of the context.

**Practices**

a) locally produced materials
A2 - Bio Pastificio IRIS (IT, 2015)
A15 - Brown Sugar Factory (CH, 2016)
B1 - UC Davis Wine Brewery (CA, 2013)
B2 - Cantina Pizzolato (IT, 2016)
B4 - Bodegas Portia (ES, 2010)

b) finishing and furnitures from local firms
B11 - Valprè (ZA, 2011)

# MR 4.1 NATURAL MATERIALS

**Type**
Technological

**Building part**
N.D.

**Impact category**
Material resources

**Material**
Wood

**Intervention**
N.D.

**Scale**
Building

**Strategy**
Renewable resources

Green marketing

**Description**

Reduce the building's environmental impact throughout its life cycle and protect workers' health by using as natural and healthy building materials as possible.

Hence materials that are not obtained from petrochemical synthesis processes, but rather from renewable resources and that have been transformed or changed at minimum (Provincia di Bologna, 2008).

According to U.S. GBC (2014) environmentally preferable attributes include: support the local economy; being sustainably grown and harvested; having intended end-of-life scenarios that avoid landfill; contain recycled content from industrial or consumer sources; being made of bio-based material; free of toxins; long-lasting, durable, and reusable.

In order to avoid green-washed products and lead the consumer to truly green products, third-party certifications and labels are spreading, also framed by policy-makers at the European level.

**Practices**

a) crosslam from FSC and natural fiber insulations
A2 - Bio Pastificio IRIS (IT, 2015)

b) biodegradable materials (naturally-treated wood)
A3 - Almazara Olisur (CL, 2009)
A6 - Solare Manufaktur Peter (DE, 2012)

c) reuse of cavity in the rock wall
A9 - Celle ipogee Melinda (IT, 2004)

d) bearing structure and cladding in wood
A23 - Carnitalia (IT, 2016)
A27 - Tangshan Organic (CH, 2016)

e) American cedar cladding without chemical treatment (for this slight distance from the ground)
A26 - Fratelli Beretta (U.S., 2016)

# MR 5.1 RECOVER OF CONSTRUCT. & DEMOLITION WASTE

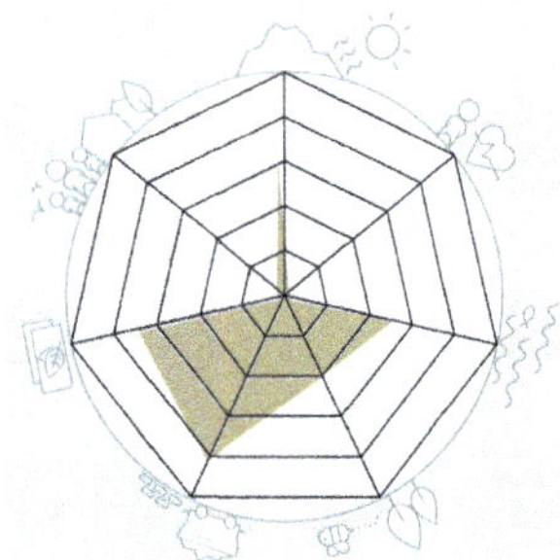

**Type**
Management

**Building part**
N.D.

**Impact category**
Material resources

**Material**
N.D.

**Intervention**
N.D.

**Scale**
Building

**Strategy**
New material saving

**Description**

Large amounts of solid waste are produced during building construction, and trash is generated throughout the building's life cycle as new items are delivered and used materials are discarded.

These wastes can be landfilled, burned, recycled, or composted. Solid waste transportation and the production of methane in landfills both contribute directly to greenhouse gas emissions (U.S. GBC, 2014).

As a result, it is critical to manage building works in such a way that some of the construction and demolition wastes (C&D) are reused, at least for earthworks (e.g., embankments, aggregates for stabilized goods and construction), hence reducing soil, transportation, and energy costs (O.P.U.Q.P., 2015).

**Practices**

a) reuse of all discards (no waste to landfill)
B13 - St. James Gate Brewery (UK, 2011)

# EQ 2.1 ECO-FRIENDLY FINISHING

**Description**

Prefer materials that emit few or no vapours, odours, powders, particles, and microfibres, as well as other dangerous and/or polluting chemicals during production, application. Select materials that emit little or no Volatile Organic Composts (VOCs), with special attention paid to solvent-based paints and adhesives (Provincia di Bologna, 2008; Kaur, Gupta, Syal, 2016).

The indoor comfort and the health of occupants will greatly benefit from this choice.

**Type**
Technological

**Building part**
N.D.

**Impact category**
Climate change

**Material**
Wood

Coating and paint

**Intervention**
N.D.

**Scale**
Building

**Strategy**
Well-being of workers

Green marketing

**Practices**

N.D.

# EQ 5.1 NATURAL VENTILATION

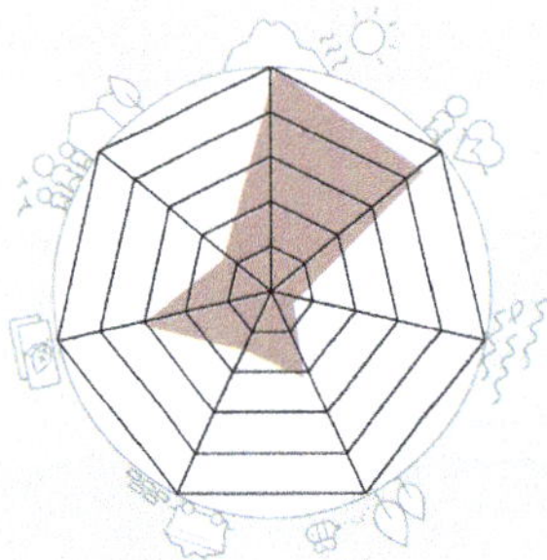

**Type**
Technological

**Building part**
N.D.

**Impact category**
Climate change

**Material**
N.A.

**Intervention**
N.D.

**Scale**
Building

**Strategy**
Bioclimatic

Well-being of workers

Green marketing

**Description**

Adopt passive strategies to improve indoor air quality and comfort through natural ventilation for benefits during summer and mid-seasons.

Natural ventilation openings must be suitably positioned in the transverse direction and elevation, in accordance with the prevailing winds and local breezes. The automatic or mechanically controlled opening systems, which help to both summer cooling and air exchange inside the rooms, play a critical role in the strategy (Provincia di Bologna, 2009). It is strongly advised to install monitoring devices to ensure that the air exchange is enough and that the amount of pollutants is within limits.

The practice of free cooling, which makes use of temperature differences between the soil and indoor spaces, where the first is usually lower in summer, is also interesting. Pipes can be installed beneath the building (on average 20cm in diameter, 10m long, and between 1.5 and 3.0m deep), allowing for the exchange of fresh air at a low cost.

Furthermore, an awareness-raising campaign among employees can be launched to encourage them to adapt to seasonality: for example, wearing a warmer dress in winter allows the company to set the internal temperature slightly lower.

**Practices**

a) free cooling

A7 - Kerrygold (UK, 2008)

b) smart system to open/close windows

A1 - Melinda (IT, 2014)

c) positioning of windows to favour air flow

A3 - Almazara Olisur (CL, 2009)

B12 - Bavaria Brewery (CO, 2011)

d) controlled opening of window + ventilator in line to exploiting thermal difference during the night

B1 - UC Davis Wine Brewery (CA, 2013)

# EQ 6.1 AUTOMATIC LIGHTING

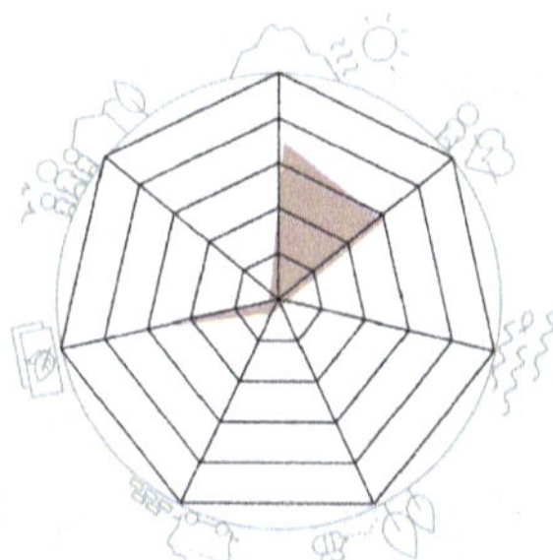

**Type**
Technological

**Building part**
Systems

**Impact category**
Climate change

**Material**
N.A.

**Intervention**
N.D.

**Scale**
Building

**Strategy**
Energy efficiency,
Visual comfort

**Description**

Small actions based on devices and sensors can increase lighting system efficiency.

Install a control system that, for instance, adjusts artificial lighting based on natural illumination levels (daylight sensors) or the presence of people (presence sensors and time switches) (Duflou et al., 2012). The system can be controlled by a "on/off" switch or by varying illumination fluxes (dimming) (Provincia di Bologna, 2008).

Daylight controls, in example, assist in lowering or shutting off electrical lights totally when daylight is sufficient; they can be zoned such that spaces near windows have lower artificial lighting levels, while those in the building's centre have higher levels of artificial light (U.S. GBC, 2014).

**Practices**

a) outside lighting sensors automatically adjust interior lights

A1 - Melinda (IT, 2014)

B13 - St. James Gate Brewery (UK, 2011)

# EQ 6.2 HIGHLY-EFFICIENT LAMPS

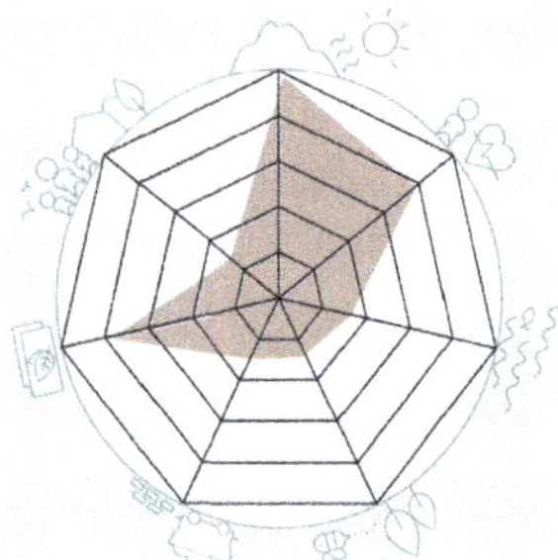

**Type**
Technological

**Building part**
Systems

**Impact category**
Climate change

**Material**
N.A.

**Intervention**
N.D.

**Scale**
Building

**Strategy**
Energy efficiency,
Visual comfort

**Description**

Make use of energy-saving lighting technologies.
LED technology, for example, promises significant energy and cost savings while simultaneously improving road visibility (from the outside) (Comune di Bomporto, 2016). Furthermore, the light quality is superior for eye health when compared to traditional neon or low-consumption lamps; they do not include dangerous components such as mercury and krypton, do not distribute part of the energy in heat, and are more resistant to constant on/off switching. Delpiano and Trombetti (2011), on the other hand, suggest discharge lamps with high frequency electronic ballasts being used instead of traditional neon lamps with electromagnetic ballasts, because they are more efficient. Additional electrical ballasts are used to regulate the light flux in the "dimming" type.

**Practices**

a) LED lights
A6 - Solare Manufaktur Peter (DE, 2012)
B11 - Valprè (ZA, 2011)
B13 - St. James Gate Brewery (UK, 2011)

# EQ 7.1 DIFFUSED TOP LIGHTING

**Type**
Technological

**Building part**
Horizontal envelope

**Impact category**
Health and well-being

**Material**
Glass

**Intervention**
New Construction

**Scale**
Building

**Strategy**
Energy efficiency,
Visual comfort

**Description**

Using as much sunlight as possible to illuminate indoor spaces (Duflou et al., 2012), which is defined as the most sustainable and comfortable lighting for humans.

Because of the depth of the rooms and the fact that they frequently store storage and machinery, the external walls of manufacturing buildings often do not ensure a sufficient supply of natural light inside the spaces. As a result, the roof provides the majority of the natural light in this form of architecture.

The traditional "shed" cover, with saw teeth facing north, allows for diffused and indirect sunlight from above, preventing glare and refraction. Avoid using direct top lighting systems (such as domes) because they might cause glare and overheating (Provincia di Bologna, 2009).

Solar tubes are also particularly interesting, as they provide diffused daylight from the top at no operational cost.

**Practices**

a) shed roof
A1 - Melinda (IT, 2014)

b) light cuts in the ceiling to reduce electricity demand
A3 - Almazara Olisur (CL, 2009)
B12 - Bavaria Brewery (CO, 2011)

c) large circular window in the ceiling
B3 - Cantina Antinori (IT, 2013)

d) solar tubes
A7 - Kerrygold (UK, 2008)

# EQ 7.2 GLASS FAÇADE

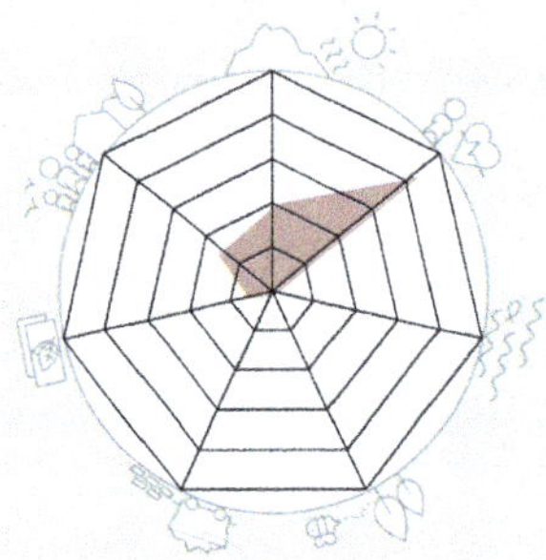

**Type**
Technological

**Building part**
Vertical envelope

**Impact category**
Health and well-being

**Material**
Glass

**Intervention**
New Construction

**Scale**
Building

**Strategy**
Visual comfort

**Description**

Provide some glass walls for the benefit of the workers' visual fatigue (quality of sunlight) and psychological well-being (quality of views).
Large glass walls are uncommon in industrial buildings today since the walls are typically used for storage and storage space; nevertheless, in cold places, large south-facing walls can provide significant benefits in terms of solar gain, reducing heating bills in the winter.
When glass façade are used, be sure that the energy performances of the element are suitable to the use (e.g., reflective or low-emitting glasses).

**Practices**

a) large glass walls to exploit solar gains
A2 - Bio Pastificio IRIS (IT, 2015)
A14 - Tofu Factory (CH, 2018)
A20 - Harim Pet Food (KR, 2016)
B6 - Dalmunach distillery (UK, 2015)
B11 - Valprè (ZA, 2011)
B13 - St. James Gate Brewery (UK, 2011)

b) ground floor entirely in glass envelope
A15 - Brown Sugar Factory (CH, 2016)

c) several windows allows workers to see the outside and benefit from daylight
A24 - Bell & Evans (U.S., ongoing)

d) large windows that brings daylight inside
B12 - Bavaria Brewery (CO, 2011)

# EQ 7.3 FILTER FAÇADE

**Type**
Technological

**Building part**
Horizontal envelope
External partition

**Impact category**
Health and well-being

**Material**
Wood
Glass
Vegetation

**Intervention**
N.D.

**Scale**
Building

**Strategy**
Energy efficiency
Visual comfort
Brand image

**Description**

When designing extensive glazing, depending on the region, take control to cover the glass from summer sun radiation, lowering heat intake during hot seasons (Provincia di Bologna, 2008).

**Physical screens that filter solar radiation, such as certain façade cladding, can be beneficial in this regard.**

The screens are also useful for preventing unwanted introspection and so protecting employees' privacy while ensuring a clear view of their surroundings and adequate lighting.

The use of micro-perforated or stretched sheet metal claddings, translucent panels, or sunshades further adds to the building's aesthetic character.

**Practices**

a) micro-perforated metal sheets to screen sun and unwanted sight
A1 - Melinda (IT, 2014)

b) bundle of sticks that shield the glass façade
A18 - Iroha Village Factory (J, 2018)

c) vertical wooden frame
A26 - Fratelli Beretta (U.S., 2016)

d) polycarbonate semi-transparent panels to avoid direct sight
A27 - Tangshan Organic (CH, 2016)

e) horizontal brise-soleil
B9 - Cantina di Collemassari (IT, 2005)

# EQ 7.4 HIGH REFLECTANCE INDOOR FINISHING

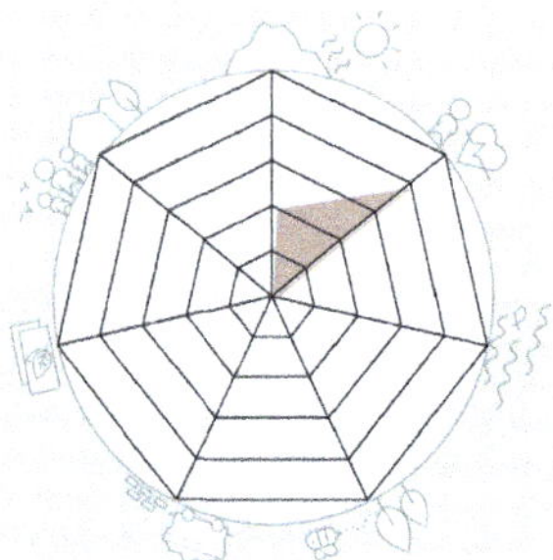

**Type**
Technological

**Building part**
Internal partition

**Impact category**
Health and well-being

**Material**
Coating and paint

**Intervention**
N.D.

**Scale**
Building

**Strategy**
Visual comfort

**Description**

Increase the indoor lighting conditions of generally dark places, such as large industrial rooms, by using materials and colours that carefully define the cladding components based on their reflectance.
However, in order to prevent negatively impacting the context, particular care should be taken to manage glare and reflection orientations.
Aside from reflecting paint and materials, suitably sized and positioned windows can significantly improve the quantity of sunshine delivered into a space. Clerestory windows, light shelves, and skylights are examples of such openings (U.S. GBC, 2014).

**Practices**

a) above-ground volumes with reflective coating increase semi-underground lighting levels
A19 - Opificio Salpi (IT, 2010)

b) light coloured finishing indoor
A24 - Bell & Evans (U.S., ongoing)

# EQ 8.1 VIEW OF THE SCENERY

**Description**

Making work settings more pleasant by placing windows that allow natural light in and allow employees to gaze out has a good impact on employees' psycho-physical well-being and, as a result, their productivity (U.S. GBC, 2014). **Provide focused views on specific landscape components,** especially if you are in a high-quality landscape context that has a restorative function (e.g. a view of trees, vegetation, rivers or lakes, mountains, but also towers and bell towers, fortresses).

**Type**
Formal

**Building part**
Vertical envelope

**Impact category**
Health and well-being

**Material**
Glass

**Intervention**
New Construction

**Scale**
Building

**Strategy**
Well-being of workers

Physical link with the context

**Practices**

a) glass wall pointing the scenery
A1 - Melinda (IT, 2014)
A2 - Bio Pastificio IRIS (IT, 2015)
A14 - Tofu Factory (CH, 2018)
A19 - Opificio Salpi (IT, 2010)
A26 - Fratelli Beretta (U.S., 2016)
B2 - Cantina Pizzolato (IT, 2016)
B6 - Dalmunach distillery (UK, 2015)
B8 - San Pellegrino (IT, ongoing)
B10 - Cantina Pieve Vecchia (IT, 2012)

b) windows facing the site green open spaces
A18 - Iroha Village Factory (J, 2018)

c) openings are positioned according to pleasant view on the scenery
A27 - Tangshan Organic (CH, 2016)

# EQ 9.1 PROTECTION FROM NOISE OF MACHINERY

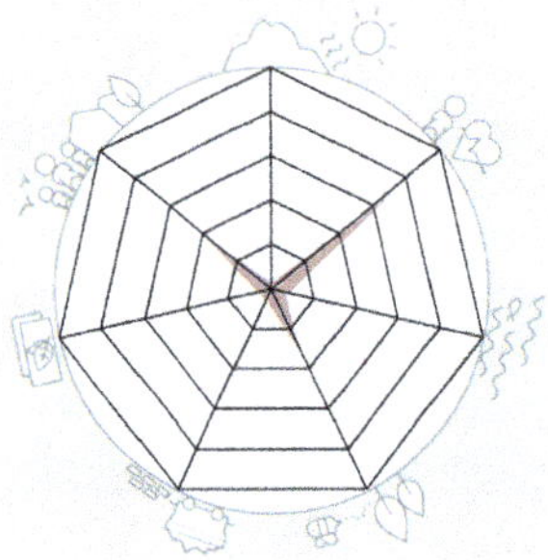

**Type**
Technological

**Building part**
N.D.

**Impact category**
Health and well-being

**Material**
N.D.

**Intervention**
N.D.

**Scale**
Building

**Strategy**
Acoustic comfort

**Description**

Controlling noise sources, containing sound reverberation, and reducing employee distraction can all be accomplished **with soundproofing techniques and equipment.**

To attain this, noisy equipment and machinery must be strategically placed in relation to sensitive units, and areas that are less exposed to noise (either internal or external) **must be designated for office, trade, and service functions** that usually must adhere to stricter acoustic requirements.

Furthermore, pay particular attention to designing construction solutions that reduce noise transmission, with a focus on nodes of structures and relative acoustic bridges (Provincia di Bologna, 2008).

**Practices**

a) sound-absorbing panels cover the interior walls
A1 - Melinda (IT, 2014)

# IN 1.1 PROCESS AUTOMATION

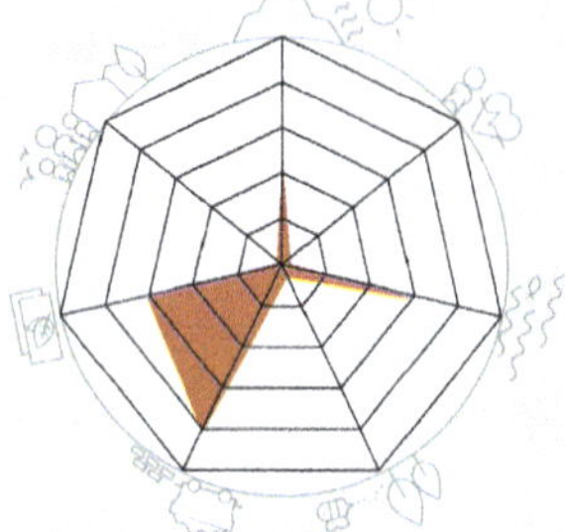

**Type**
Technological

**Building part**
Systems

**Impact category**
Material resources

**Material**
N.A.

**Intervention**
N.D.

**Scale**
Process

**Strategy**
Automatic system

**Description**

**Making the production process more efficient by automation** it entails shifting from low productivity "labour intensive" logistics to high productivity "technology intensive" logistics (Aster, 2014).

In general, this involves the installation of automated systems for handling, storage, order preparation, sorting, and technologies like augmented reality, as well as the usage of ICT (Information and Communication Technologies) (May, Stahl and Taish, 2016).

**As specific practice,** Delpiano and Trombetti (2011) recommend adopting variable-speed electric motors to conserve energy based on load: instead of using throttle valves and dampers, these also allow for soft starting and raising the power factor.

**Practices**

a) information technology that reduces production errors and control indoor comfort
A2 - Bio Pastificio IRIS (IT, 2015)

b) fully automated plant to save energy and water
A20 - Harim Pet Food (KR, 2016)
A24 - Bell & Evans (U.S., ongoing)

c) advanced robotics to move ingredients based on flexible demand
A21 - Van Gelder (NL, 2019)

d) information technology for quality control and security of the building
A23 - Carnitalia (IT, 2016)

e) conveyor belt with "soft start"
B11 - Valprè (ZA, 2011)

# IN 1.2

# RECOVER HEAT / ENERGY FROM PROCESSES

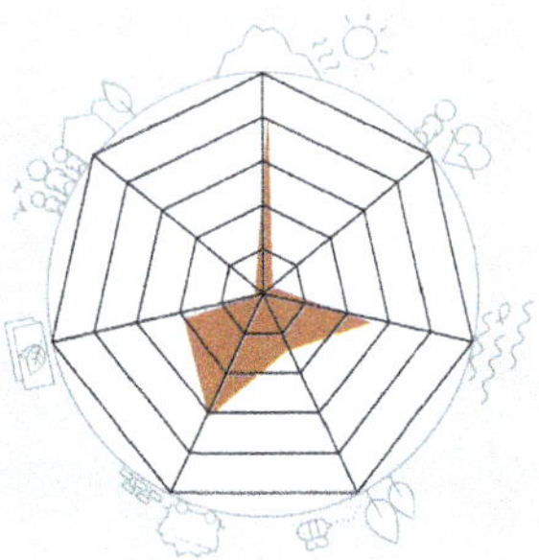

**Type**
Technological

**Building part**
Systems

**Impact category**
Climate change

**Material**
N.A.

**Intervention**
N.D.

**Scale**
Process

**Strategy**
Energy efficiency

Green marketing

**Description**

Try to recover heat or other forms of energy from industrial operations of companies that produce liquid or gaseous effluents in sufficient quantities to power energy systems serving the company and other businesses in the area (Provincia di Bologna, 2008).
For instance, residual heat from air compressors can be used to pre-heat water. Hot air at the end of a production cycle can be transferred to an exchanger to assist with the building winter heating. In advance, simulation tools can be used to analyse the effectiveness and efficiency of these connections (Ball et al., 2013).
Due to the relatively low operating temperatures of the processes (40-140 °C), heat recovery from manufacturing processes is not commonly used in the food sector, but Miah et al. (2015) propose accurate analysis of the system to detect unexpected but valuable residual energy/heat also in this field.

**Practices**

a) recover of heat from refrigerators and compressors to warm the building A1 - Melinda (IT, 2014)

b) pre-heating hot water (55% of energy use) by residual heat A24 - Bell & Evans (U.S., ongoing); B7 - Carlsberg (FR, 2014)

c) heating the visitor centre through residual heat B5 - Macallan distillery (UK, 2018); B14 - Bombay Sapphire Distillery (UK, 2013)

d) residual heat reused in the manufacturing cycle B6 - Dalmunach distillery (UK, 2015)

e) recover of compressed air to use in the process (bottle blowing) B11 - Valprè (ZA, 2011)

f) residual process energy reused to limit steam production B13 - St. James Gate Brewery (UK, 2011)

# IN 1.3 ENERGY FROM WASTE STREAMS

**Type**
Technological

**Building part**
Systems

**Impact category**
Climate change

**Material**
N.A.

**Intervention**
N.D.

**Scale**
Process

**Strategy**
New material saving
Green marketing

**Description**

Recover energy from waste processing to make a company more environmentally friendly in general. Organic by-products, for example, can be re-used for biogas production by the same company or another firm that is part of a circular economy process.

The BREEAM certification procedure does not positively acknowledge first generation biogas - that is, biogas produced with organic fuels explicitly produced for this purpose - but it does support the recovery and transformation of production waste and second-generation biogas products (BRE, 2016).

The food industry is particularly interested by the topic. Despeisse (2013), for example, mentions a case of anaerobic treatment of organic ingredients in a brewery's process water to generate methane, as well as a case of a sugar cane processing company.

**Practices**

a) co-generation plant and biogas production from organic waste
A11 - Amadori (IT, 2016)

b) energy from biomass of production waste
A17 - Frito-Lay (TX, 2011)
B14 - Bombay Sapphire Distillery (UK, 2013)

c) biogas from water sludge
B7 - Carlsberg (FR, 2014)

# IN 1.4 TREATMENT AND REUSE OF GREY WATER

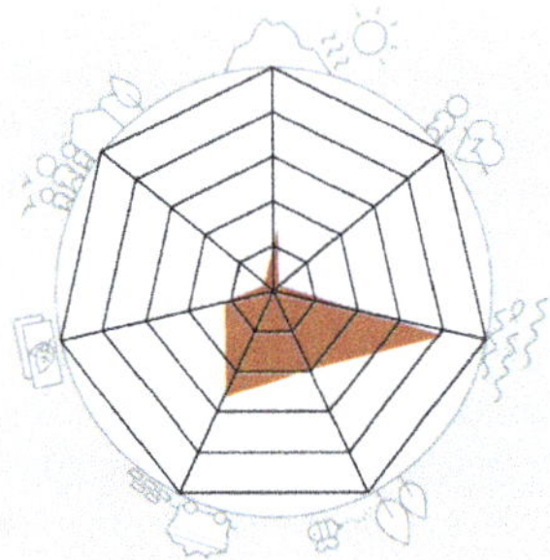

**Type**
Technological

**Building part**
Systems

**Impact category**
Water resource

**Material**
N.A.

**Intervention**
N.D.

**Scale**
Process

**Strategy**
Water saving

**Description**

Treat grey water with the goal of reusing it for secondary purposes such as irrigation of gardens, lawns, urban uses, or feeding ponds and wetlands by reintroduction into the aquifer or water channels (Provincia di Milano, 2012).

Natural systems such as phyto-depuration can clean rainwater, municipal waste, and a portion of production water while also benefiting the landscape and site attractiveness (O.P.U.Q.P., 2015): a stratified substrate with varying soils and the planting of appropriate plant species purifies water while protecting native habitats and wildlife diversity (Grant, 1997).

As an example of a wastewater treatment and water recycling plant, DEFRA (2006) provides an advanced membrane bio-reactor and reverse osmosis technology (namely AMBR). It is an intensive aerobic biological process that uses a water separation membrane instead of typical sedimentation processes. The use of AMBR in a case study resulted in a considerable reduction in effluent discharge volume, well within the discharge limitations.

Moreover, industrial waste water can follow sustainable energy processes, as recovering residual heat.

**Practices**

a) treated process water fills decorative ponds, and dark water is treated by plant-based systems
A2 - Bio Pastificio IRIS (IT, 2015)

b) plant-based system to treat dark water
B1 - UC Davis Wine Brewery (CA, 2013)
B2 - Cantina Pizzolato (IT, 2016)

c) treatment of grey water and partial reuse
B11 - Valprè (ZA, 2011)

# IN 1.5 TREATMENT AND REUSE OF PROCESS WATER

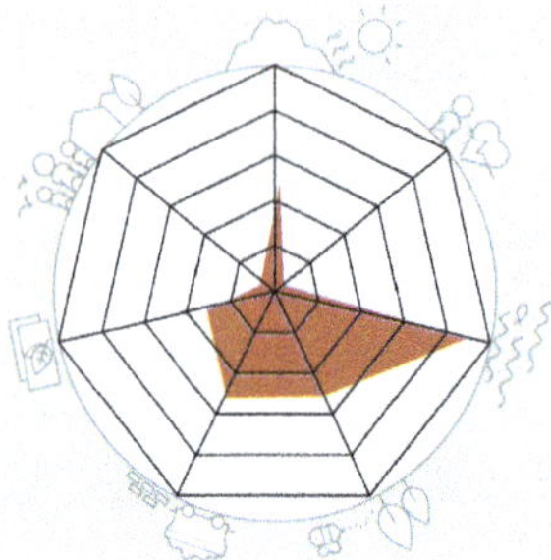

**Type**
Technological

**Building part**
Systems

**Impact category**
Water resource

**Material**
N.A.

**Intervention**
N.D.

**Scale**
Process

**Strategy**
Water saving

**Description**

Adopt water-saving measures in the industrial sector, where the concept of "fit for use," or using water with a quality no higher than that required for the use to which it is put, is gaining traction.

It is easier to recover and reuse process water when it is kept as distinct as possible according to the need of contaminant or treatment homogeneity.

In fact, segregating contaminants by category makes it simpler to eliminate them, avoiding the need to treat the entire flow of water or just discharge them into the water body (Aster, 2014).

**Practices**

a) water cooling cycle is integrated with basins for neighbourhood's water landscaping
A9 - Celle ipogee Melinda (IT, 2004)

b) in-house technology for treating the water
A11 - Amadori (IT, 2016)

c) filters with semi-permeable membranes allow water to be used up to 10 washing cycles
B1 - UC Davis Wine Brewery (CA, 2013)

d) inverse-osmosis to reuse production water
B5 - Macallan distillery (UK, 2018)

e) water electrochemically activated is used in Clean-In-Place (CIP) processes free from chemicals
B11 - Valprè (ZA, 2011)

# IN 1.6 TECHNOLOGY TO ABATE POLLUTION

**Type**
Technological

**Building part**
N.D.

**Impact category**
Climate change

**Material**
Concrete

Coating and paint

**Intervention**
N.D.

**Scale**
Building

**Strategy**
Air pollution reduction

Green marketing

**Description**

Reduce industrial $CO_2$ emissions by implementing one of the many techniques now available. End-of-pipe filters/technology, which capture pollutants at the end of the process, are the most widely used and cheapest (Despeisse et al., 2017).
Others, which are more sophisticated and effective in this regard, concentrate on energy-saving techniques taken throughout the production process, or even preventive actions implemented upstream.
In terms of minimizing emissions from the industrial building itself, choose construction materials with low environmental impact during manufacturing and disposal, as well as new materials that aid to reduce organic pollution deposited on the building's surfaces (e.g., coatings with titanium).
To this end, Herrmann et al. (2015) suggest the use of novel materials that can absorb pollutants or cladding systems with built-in photovoltaic for electricity generation. Not to mention Nature-based solutions such as a green roof or façade that can give a positive contribution to the problem.

**Practices**

a) coating with titanium which catalyses organic pollutant decomposition
A2 - Bio Pastificio IRIS (IT, 2015)

b) use technology that retains $CO_2$
A9 - Celle ipogee Melinda (IT, 2004)
B1 - UC Davis Wine Brewery (CA, 2013)

c) green roof to fix particulate matter
A19 - Opificio Salpi (IT, 2010)

# RP 1.1 SHARING OF FACILITIES IN CONSORTIUM

**Type**
Management

**Building part**
N.D.

**Impact category**
Material resources

**Material**
N.A.

**Intervention**
N.D.

**Scale**
Site

**Strategy**
Energy efficiency
Green marketing

**Description**

For small artisan enterprises that cannot afford it, sharing the energy production system or investing in a new and more efficient central heating system can be an economic and energy-saving option.
In some circumstances, spatial proximity and a shared set of goals with near businesses may induce industrial symbiosis or cycle closure processes.

**Practices**

a) sharing of facilities and storage spaces as consortium
A14 - Tofu Factory (CH, 2018)

b) close cycle: waste plastic is reused by another local company
B11 - Valprè (ZA, 2011)

# RP 1.2 SHARING OF SERVICES AND MIXED USE

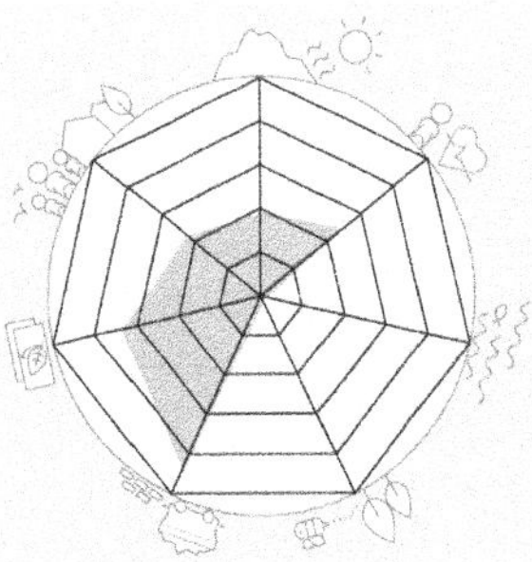

**Type**
Management

**Building part**
N.D.

**Impact category**
Material resources

**Material**
N.A.

**Intervention**
New construction

**Scale**
Site

**Strategy**
New material saving
Community engagement

**Description**

Develop production areas that are linked to services, infrastructure, and common public spaces based on the eco-industrial parks concept (O.P.U.Q.P., 2015).

This means that small businesses that lack the financial resources to cover the costs of employee spaces and services, such as a canteen or a kindergarten for employees' children, can share spaces and infrastructure with other nearby businesses, allowing workers to access services that would otherwise be unavailable.

This, in turn, contributes to increasing the range of uses in a specific area and attracting people other than workers to interact with and walk through the site.

**Practices**

N.D.

# PA 1.1 VISUAL SCREEN

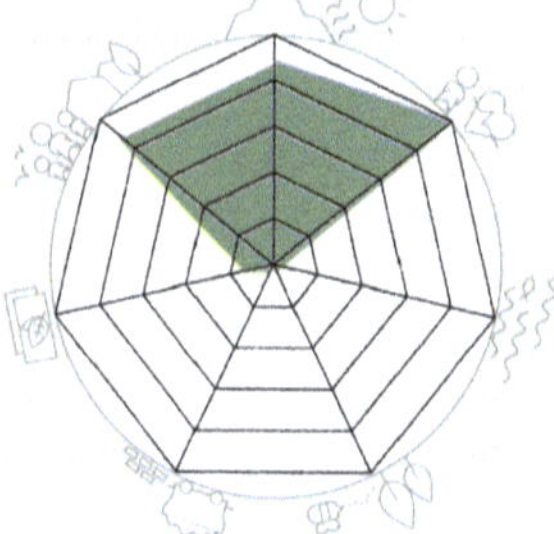

**Type**
Formal

**Building part**
Vertical envelope
Internal partition

**Impact category**
Health and well-being

**Material**
N.D.

**Intervention**
N.D.

**Scale**
Building

**Strategy**
Energy efficiency
Visual comfort
Visual mitigation

**Description**

Create visual screens to minimize building visibility or to harmonize different volumes. Since the early design, consider the potential points of view (e.g., eye height, as seen from an elevated road or from a hill) and the type of perception that observers will have (e.g., static or dynamic). Tandy (1975) identifies two types of natural shielding: terrain and vegetation. Because any additional "vegetation" on an industrial scale may turn out to be an artefact, the approach recommends introducing plant screens that are the same size and features as what already exists. On the same wavelength, O.P.U.Q.P. (2015) advises grouping trees organically rather than in line to achieve a more "natural" image.

Alternatively, screens can also be "technological". It comes, for instance, of a building's second skin, such as a ventilated façade, or a standalone structure with the third dimension. As a result, it also helps with protection of the building from atmospheric agents (Dessì et al., 2018) and with architectural and chromatic integration of the site in the scenery (Herrmann et al., 2015). Screens in fact can also modify the perception of volumes, for example by de-structuring the mass towards the roof.

**Practices**

a) screen that visually de-constructs as it goes up
A3 - Almazara Olisur (CL, 2009)

b) inclined wood joists remind of surrounding forests
A8 - Pastificio Felicetti (IT, 2014)

c) a continuous wooden curtain become a filter wall at the windows A26 - Fratelli Beretta (U.S., 2016)

d) wooden grate uniform the skyline and visually mitigate service area B2 - Cantina Pizzolato (IT, 2016)

e) perforated metal sheets to hide the silos and service area in the high quality scenery
B15 - CE.VI.V. (IT, 2018)

# PA 1.2 HORIZONTAL DIMENSION

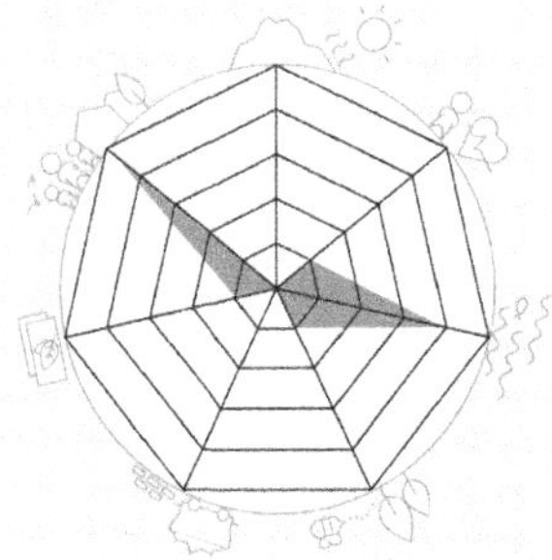

**Type**
Formal

**Building part**
N.D.

**Impact category**
Community

**Material**
N.A.

**Intervention**
New construction

**Scale**
Building

**Strategy**
Visual mitigation

**Description**

Make the horizontal dimension of the building dominate over the vertical dimension to reduce the visual impact.
Industrial structures rarely surpass two floors due to man-based logistical limits that make preferring industry development per horizontal surface. However, the recent advent of completely automated warehouses, as well as efforts to limit soil consumption by reducing the building footprint, have pushed businesses to develop new buildings by height.
Therefore, the O.P.U.Q.P. (2015) advises against locating industrial structures on ridges and avoiding heights that provide a visual barrier from the primary observation channels, particularly in the presence of hilly landscapes. If at all possible, build an underground part to lessen the site's visual impact.

**Practices**

a) development of the complex in length
A3 - Almazara Olisur (CL, 2009); B8 - San Pellegrino (IT, ongoing)

b) short but lengthened building as the nearby forest, slightly raised on the ground to visually suspend it
A26 - Fratelli Beretta (U.S., 2016)

c) short but lengthened building to connect with the surrounding plain surface
A27 - Tangshan Organic Farm (CH, 2016)

d) height limiting and vertical differentiation of materials make the building perceived as lower
A28 - Mulino Agri'90 (IT, 2011); B12 - Bavaria Brewery Tocancipá (CO, 2011)

e) underground construction with one storey above ground
B10 - Cantina Pieve Vecchia (IT, 2012)

# PA 1.3 LANDFORM

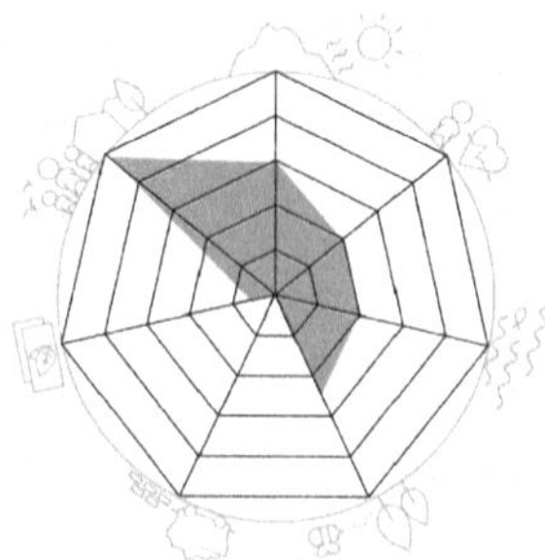

**Type**
Formal

**Building part**
Outdoor equipment

**Impact category**
Community

**Material**
Vegetation and/or water

**Intervention**
New construction

**Scale**
Site

**Strategy**
Bioclimatic,
Visual mitigation

**Description**

Industrial facilities can be visually mitigated by landform design (Tandy, 1975).

Appropriate planning of earthwork allows for mechanisms of morphological integration by using excavation soil and rocks from the building site. For example, slight modifications of the terrain profile can improve the perception of the construction, by partially hiding the view of its less pleasant parts (e.g. silos, service areas) or rather by attracting attention to the most significant elements (Busquets i Fabregas, 2007).

The tactic includes additive (hills, hillocks) and subtractive interventions (land depression). The latter is especially useful when huge buildings are located in close proximity to natural elements, as the need of visual screen is reduced and good scenery perspectives are ensured (O.P.U.Q.P., 2015).

**Practices**

a) complex adapts to the slope
A14 - Tofu Factory (CH, 2018)

b) ground depression to reduce visibility and to facilitate operations of the production process
B4 - Bodegas Portia (ES, 2010)

c) green hill to reduce visibility
B5 - Macallan distillery (UK, 2018)
B6 - Dalmunach distillery (UK, 2015)

# PA 1.4 UNDERGROUND BUILDING

**Type**
Formal

**Building part**
N.D.

**Impact category**
Climate change

**Material**
Vegetation a/o water

**Intervention**
New construction

**Scale**
Building

**Strategy**
Bioclimatic

Visual mitigation

**Description**

Making all or part of a building below ground level can **save land, limit factory height, and benefit from ground** thermal storage in both cold and hot seasons.

Indeed, the constant thermal temperature of soil has long been used for food preservation in both the commercial and residential sectors, and it appears to have been rediscovered by some successful companies that use subterranean cold storage, both for the associated energy savings and for the **aesthetic mitigating benefits.**

**Because food firms are commonly located in rural or high-**quality landscape settings, the design of an underground **structure may help to significantly lessen the perceived** impact.

**Practices**

a) green hill + underground building
A19 - Opificio Salpi (IT, 2010)

b) only the entrance of the warehouse is above ground
A9 - Celle ipogee Melinda (IT, 2004)

c) the building is as like a fence in the landscape
B3 - Cantina Antinori (IT, 2013)

**d) underground facilities and above ground offices**
B9 - Cantina di Collemassari (IT, 2005)
B10 - Cantina Pieve Vecchia (IT, 2012)

# PA 1.5 REFERENCE TO LOCAL CONSTRUCTIONS

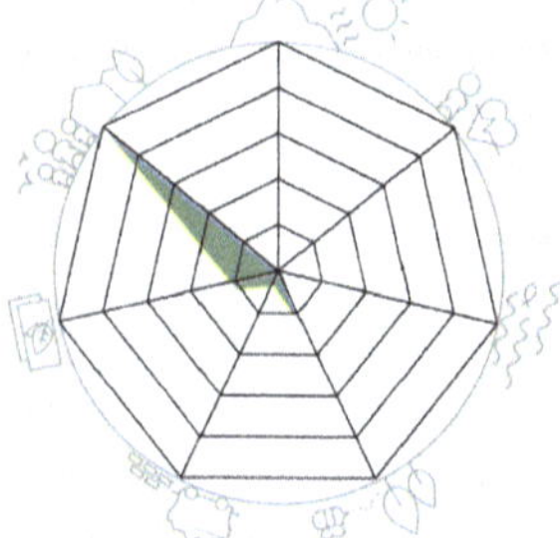

**Type**
Formal

**Building part**
Top horizontal envelope
Vertical envelope

**Impact category**
Community

**Material**
N.A.

**Intervention**
New construction

**Scale**
Building

**Strategy**
Physical link with the context

**Description**

Take as a reference the buildings or construction elements that are considered "typical". That are those rooted in construction traditions and local imagery, and adhere as closely as possible to the compositional rules of the urban texture. The tactic aim at avoiding too striking contrasts of the project with the territory.

**Practices**

a) reference to building with court as traditional type
A27 - Tangshan Organic (CH, 2016)

b) reference to the shape of traditional houses
B6 - Dalmunach distillery (UK, 2015)

c) reference to the arch as key element of local historic buildings
B8 - San Pellegrino (IT, ongoing)

# PA 1.6 REFERENCE TO THE SKYLINE

**Type**
Formal

**Building part**
Top horizontal envelope
Vertical envelope

**Impact category**
Community

**Material**
N.A.

**Intervention**
New construction

**Scale**
Building

**Strategy**
Physical link with the context

**Description**

Respect the character of the place and avoid going against the local landscape characteristic, features and forms.
Then perform a site assessment by examining the primary lines of the scenario (e.g. hills, mountains) and try to maintain the accordance with them.

**Practices**

a) slope lines of the roof refers to the mountains in the skyline
A22 - Caffè Milani (IT, 2017)

b) the building shape refers to lines of natural elements in the panorama
B5 - Macallan distillery (UK, 2018)
B10 - Cantina Pieve Vecchia (IT, 2012)

# PA 1.7 VOLUME BREAKDOWN

**Type**
Formal

**Building part**
N.D.

**Impact category**
Community

**Material**
N.D.

**Intervention**
New construction

**Scale**
Building

**Strategy**
Visual mitigation

**Description**

Compositional techniques such as articulation of the volume into parts, architectural elements, openings, coloured background, and so on can help to reduce the visual impact of oversized volumes.

In areas with fine-grained urban fabric, a compositional solution of smaller aggregate volumes is preferred over a single huge building that may be out of scale. The functional organization might help with morphological subdivision of the site, which has advantages in terms of differentiating uses (e.g. day/night uses) and performance (e.g. different comfort requirements) for distinct areas.

In addition, instead of establishing continuous, monotonous fronts, leave small openings toward the horizon (O.P.U.Q.P., 2015). Therefore break up massive facade surfaces using constructive elements (windows, scores) or natural elements to improve the perception of large blind, faceless, and monotonous volumes (trees, climbing plants).

**Practices**

a) formal division follows spaces functional separation

A14 - Tofu Factory (CH, 2018)

A15 - Brown Sugar Factory (CH, 2016)

A28 - Mulino Agri '90 (IT, 2011)

b) four diverse buildings are unified by the pathways and the court

A27 - Tangshan Organic (CH, 2016)

c) single production areas are connected by an overhanging roof

B6 - Dalmunach distillery (UK, 2015)

# PA 1.8 CLEAR LAYOUT

**Type**
Formal

**Building part**
N.D.

**Impact category**
Community

**Material**
N.D.

**Intervention**
N.D.

**Scale**
Site
Building

**Strategy**
Brand image

**Description**

Design manufacturing settlements with a compact plan concentrating the buildings around a central core or placing them on the area's main axes to avoid dispersion effects to the rural area (O.P.U.Q.P., 2015).
Refer to the settlement logic, road system, and distribution of pre-existing green areas as harmonious formal guidelines as much as feasible, and incorporate existing natural features into the concept (Busquets i Fabregas, 2007).
The layout and character of the buildings and open spaces (including trees and fences) contribute to the creation of a consistent and distinctive image of the site, both in the design of new plants and in the rehabilitation of existing areas. It is helpful to promote a coherent image of the system in order to establish a recognized architectural sign in the landscape, especially when the proportions make mitigation impossible.
The O.P.U.Q.P. recommends using coordinated materials in outdoor spaces and foliage to screen the services areas, which are typically positioned in the backwards; employ symbolic elements (furniture and installations), unifying signs (e.g., colours and materials), and visual focus.

**Practices**

a) a white frame unifies volumes with different functions
A21 - Van Gelder (NL, 2019)

b) parking and services area hidden backward
A26 - Fratelli Beretta (U.S., 2016); A28 - Mulino Agri '90 (IT, 2011)

c) services and ancillary spaces are hidden by slopes
B5 - Macallan distillery (UK, 2018)

d) unifying visual screen B2 - Cantina Pizzolato (IT, 2016); B15 - CE.VI.V. (IT, 2018)

e) new buildings are coloured in black and white to unify the complex B12 - Bavaria Brewery (CO, 2011)

# PA 1.9 KEMPT BORDERS

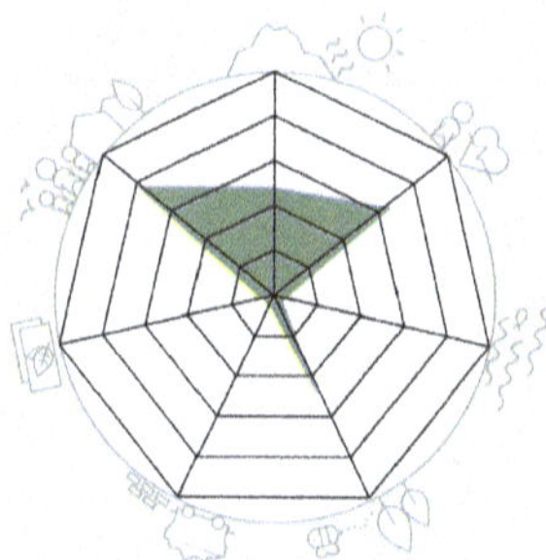

**Type**
Formal

**Building part**
Outdoor equipment

**Impact category**
Community

**Material**
Vegetation a/o water

**Intervention**
N.D.

**Scale**
Site

**Strategy**
Brand image

**Description**

Take care of the site margins to create a unified and ordered image.

The O.P.U.Q.P. (2015) recommends using plant elements (such as embankments, drainage channels, trees and bushes) rather than solid masonry or metal fences, also for ensuring the continuity of ecological networks.

Avoid planting trees in straight lines, which emphasize the border's linearity.

In urban locations, integrate with the pre-existing built fabric to recall its sense of "street," for example, positioning accesses to more permeable functions (such as offices) on the street side and service functions inside or on the back of the lot (e.g., parking).

**Practices**

N.D.

# PA 1.10 HIDDEN SETTING

**Type**
Formal

**Building part**
N.D.

**Impact category**
Community

**Material**
N.A.

**Intervention**
New construction

**Scale**
Site

**Strategy**
Visual mitigation

**Description**

Place the new building as close to natural elements as possible to reduce the visual impact on the landscape (O.P.U.Q.P., 2015). For example, next to a terrain relief.
Avoid placing the industrial complex on the top of a ridge in the presence of hilly or mountainous profiles, and avoid heights that may impede the primary landscape observation channels.

**Practices**

N.D.

# PA 2.1 COLOUR FROM THE CONTEXT

**Type**
Formal

**Building part**
Vertical envelope

**Impact category**
Community

**Material**
N.D.

**Intervention**
N.D.

**Scale**
Building

**Strategy**
Visual mitigation

**Description**

Reduce the site visual impact by strategically using colour to get the desired effect, whether it is mimesis, contrast, or singularization.

The tactic is based on a thorough analysis of the area's palette, as well as the type of plant that are in place and their seasonal colour variations (Busquets i Fabregas, 2007). On this basis, a colour palette for the project can be defined and used consistently.

If camouflage is desired, the same colour ranges should be employed. Instead, if contrast and singularization are preferred, complementary colours can be used. Green tones must be chosen with care, as they risk causing negative landscape impacts if they differ from those of the existing vegetation.

Use reflective surfaces judiciously and carefully check their effects on the environment: highly reflecting surfaces (e.g. shiny, mirrored, very clear) can induce glare effects that cause context disturbances, such as obstructing the visibility of cars, cyclists, and other road users.

**Practices**

a) cladding panels in natural tones from the surrounding
A1 - Melinda (IT, 2014)
A3 - Almazara Olisur (CL, 2009)
B3 - Cantina Antinori (IT, 2013)
B4 - Bodegas Portia (ES, 2010)

b) fruit trees and shrubs change colour during the seasons
A21 - Van Gelder (NL, 2019)

c) colours of wood and vegetation in the background
A26 - Fratelli Beretta (U.S., 2016)

d) a picture of the sky on the façade reduces the visual impact in days with the same weather of the photo
B7 - Carlsberg (FR, 2014)

# PA 2.2 PATTERN FROM THE CONTEXT

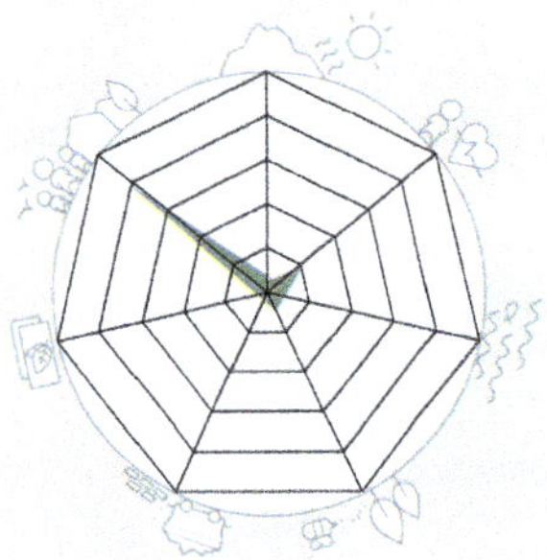

**Type**
Formal

**Building part**
Vertical envelope

**Impact category**
Community

**Material**
Wood, Vegetation, Coating and paint

**Intervention**
N.D.

**Scale**
Building

**Strategy**
Physical link with the context

**Description**

When integrating the complex into the landscape, consider the lines that define the setting, especially where these are well-known and based both in the structure of the building and in the public imagination.

The Roman centuriatio, which can still be seen in road infrastructures and land plot subdivisions in the Po Valley today, or the reference to the geological stratification of banks and escarpments near the complex, or the stylized reference to the shapes of the surrounding forest are just a few examples.

**Practices**

a) cladding panels with colours subdivisions that refers to cultivated fields
A3 - Almazara Olisur (CL, 2009)

b) sketchy reference to local trees
A8 - Pastificio Felicetti (IT, 2014)

c) reference to the ground geological stratification
B8 - San Pellegrino (IT, ongoing)
B15 - CE.VI.V. (IT, 2018)

# PA 2.3 VEGETATED ROOF

**Type**
Formal

**Building part**
Top horizontal envelope

**Impact category**
Climate change

**Material**
Wood, Vegetation,
Coating and paint

**Intervention**
New construction

**Scale**
Building

**Strategy**
Bioclimatic
Visual mitigation

**Description**

Using vegetation material to create roofing improves roof insulation, reduces rainwater run-off, and improves the building's landscape integration (Provincia di Milano, 2012).

The green roof is especially effective in rural or peri-urban areas where greenery dominates the landscape. The type of integration that can be accomplished is determined by the morphology of the building: if the shape is rigid and very high, the landscape harmonization is limited to the site view from above; if, on the other hand, the building is characterized by fluid forms that blend with the ground, the integration will be effective from a human perspective as well.

Gravel coverage is another option that help to increase the environmental performance and landscape integration of the site (O.P.U.Q.P., 2015).

**Practices**

a) extensive green roof
A1 - Melinda (IT, 2014)
A18 - Iroha Village Factory (J, 2018)
A19 - Opificio Salpi (IT, 2010)
B5 - Macallan distillery (UK, 2018)

b) intensive green roof (vineyard linked with surrounding fields)
B3 - Cantina Antinori (IT, 2013)

# PA 2.4 MATERIALS FROM THE CONTEXT

**Type**
Formal

**Building part**
Vertical envelope
Top horizontal envelope

**Impact category**
Community

**Material**
Vegetation a/o water

**Intervention**
N.D.

**Scale**
Building

**Strategy**
Visual mitigation

**Description**

Use materials that are relevant of the scenario to achieve a sense of harmony between the building and its surroundings (Provincia di Milano, 2012).
This could imply, for example, employing wood siding if near wooded areas, or using a mirror coating in some (limited) portion of the building to **reflect the surroundings.** The latter, on the other hand, necessitates a considerable attention and control during the design stage in order to **avoid the mirrored surfaces causing distracting glare effects** in the surrounding environment.

**Practices**

a) materials refers to poor traditional techniques
A15 - Brown Sugar Factory (CH, 2016)

b) mirrored cladding reflects the surrounding nature
A19 - Opificio Salpi (IT, 2010)

c) wood and mirrored glass connect with surrounding nature aiming at being seamless
A26 - Fratelli Beretta (U.S., 2016)

d) local materials re-purposed in an innovative way
A28 - Mulino Agri '90 (IT, 2011)
B2 - Cantina Pizzolato (IT, 2016)

# PA 2.5 GRAPHIC DESIGN

**Type**
Formal

**Building part**
Vertical envelope

**Impact category**
Community

**Material**
Coating and paint

**Intervention**
N.D.

**Scale**
Building

**Strategy**
Brand image

**Description**

Using chromatic and graphic techniques to reduce the visual impact of existing buildings that are out of scale. As a matter of fact, an interesting graphic/chromatic treatment of the facade can transform a monotonous and unattractive box into a pleasant, sculptural and articulated structure (Clark, 2005).

According to the colour theory (Bell, 2004):

i) fading a grey or blue colour upwards reduces the visual mass of the building, blending it in with the sky;

ii) warm tones advance and thus affirm themselves as dominant;

iii) cold temperatures cause the opposite effect of distancing and are associated with something that is far.

Moreover, colours with high brightness tend to stand out, and very bright surfaces are more appealing than dark ones (García et al., 2003). Thus, to reduce the visual mass of a building it is possible to attribute a darker colour to the roof than to the walls and to use bright tones only in small parts of the building (e.g., in the openings) (O.P.U.Q.P., 2015).

Finally, select colour treatments that have also a good impact on the environment, such as 'cool' or anti-radiation reflective coatings (Comune di Bomporto, 2016).

**Practices**

a) cladding panels in dark brown that visually reduce the factory compared to the brighter offices

A26 - Fratelli Beretta (U.S., 2016)

b) perforated metal sheets with different diameters to produce dynamic effects on the façade and de-structuring the volume

B15 - CE.VI.V. (IT, 2018)

# PA 3.1 TERRAIN OR VEGETATION BARRIER

**Type**
Technological

**Building part**
Outdoor equipment

**Impact category**
Health and Well-being

**Material**
Vegetation a/o water

**Intervention**
N.D.

**Scale**
Site

**Strategy**
Ecosystem protection
Acoustic comfort
Visual mitigation

**Description**

The use of a large plant barrier can assist in the reduction of noise generated by automobiles on a road or, conversely, from an industrial polygon.

Natural acoustic barriers may include embankments or terracing with plants and trees planted on top. Plants with dense foliage, particularly those with evergreen broad leaves, must be selected (Provincia di Milano, 2012).

However, in order to obtain a noise level reduction of more than 10 decibels, a tree barrier depth of at least 20-30 meters is required, making the method challenging to apply in an urban area.

**Practices**

a) vegetated roof to abate noise from external sources
A19 - Opificio Salpi (IT, 2010)

b) a terrain relief between the building and the road abates noise pollution from the road
B6 - Dalmunach distillery (UK, 2015)

# PA 3.2 TECHNOLOGICAL BARRIER

**Type**
Technological

**Building part**
Outdoor equipment

**Impact category**
Health and Well-being

**Material**
N.D.

**Intervention**
N.D.

**Scale**
Site

**Strategy**
Acoustic comfort
Visual mitigation

**Description**

Reduce traffic noise from a road or, in this case, an industrial polygon by building sound-absorbing barriers that can later be covered by adequate shrub and tree bands (Provincia di Milano, 2012), or possibly integrated with photovoltaic systems or other helpful devices.
As a result, this can contribute in the upgrading of the site's visual appearance.

**Practices**

N.D.

# PA 4.1 MULTIFUNCTION FACILITY

**Type**
Functional

**Building part**
N.A.

**Impact category**
Community

**Material**
N.A.

**Intervention**
New construction

**Scale**
Site

**Strategy**
Community engagement
Green marketing

**Description**

The presence of multiple functions (mix productive, sociocultural, commercial, company-related, or even external activities) on the site can promote its receptivity (Ginelli and Daglio, 2016). Busquets i Fabregas (2007) defines it as "business outsourcing", that is improving the area's quality and architectural, environmental, and landscape design and providing services that improve the quality of life in the industrial complex. In this regard, the activities are often planned to attract young people while preventing vandalism. Climbing walls or skate parks on the roof are two good examples.
The strategy is inspired by Bjarke Ingels "hedonistic sustainability," according to which sustainability should not be associated with renunciation and sacrifice, but rather with enjoyment. Therefore, ensure the accessibility and usability of the open spaces, as well as the integration of recreational and social functions open not only to employees but also to residents.
Refer to Dessì et al. (2018) who provide an overview of practices to make a public area appealing.

**Practices**

a) kindergarten and public park
A2 - Bio Pastificio IRIS (IT, 2015)
b) square, offices and food tasting space A14 - Tofu Factory (CH, 2018); B15 - CE.VI.V. (IT, 2018)
c) community centre
A15 - Brown Sugar Factory (CH, 2016)
d) research centre, cafeteria and shops
A18 - Iroha Village Factory (J, 2018)
e) visitors centre, offices, auditorium, shops
A20 - Harim Pet Food (KR, 2016); B17 - Distilleria Zanin (IT, 2013)
f) offices, museum, restaurant, educational spaces A21 - Van Gelder (NL, 2019); A22 - Caffè Milani (IT, 2017)
g) guest house
A26 - Fratelli Beretta (U.S., 2016)
h) shops and events B2 - Cantina Pizzolato (IT, 2016); B4 - Bodegas Portia (ES, 2010)

# PA 4.2 TERRITORIAL PROMOTION & SOCIAL RESPONSIBILITY

**Type**
Management

**Building part**
N.A.

**Impact category**
Community

**Material**
N.A.

**Intervention**
N.D.

**Scale**
N.A.

**Strategy**
Community engagement
Green marketing

**Description**

Create an image of the place and promote its excellence through promotion and communication initiatives (Cavallo, Degli Esposti, Konstantinou, 2012).

This includes Corporate Social Responsibility (CSR) activities, such as the adoption of ethical behaviours both internally (e.g., human resource management, worker health and safety, natural resource management, and environmental sustainability) and externally (relationships with the local community, economic partners, suppliers, customers, consumers, and respect for human rights throughout the production chain) (Provincia di Bologna, 2008).

**Practices**

a) the firm supports local economy and workers
A15 - Brown Sugar Factory (CH, 2016)

b) events for promoting the territorial food and beverage branding
B9 - Cantina di Collemassari (IT, 2005)

# PA 4.3 PUBLIC PARK

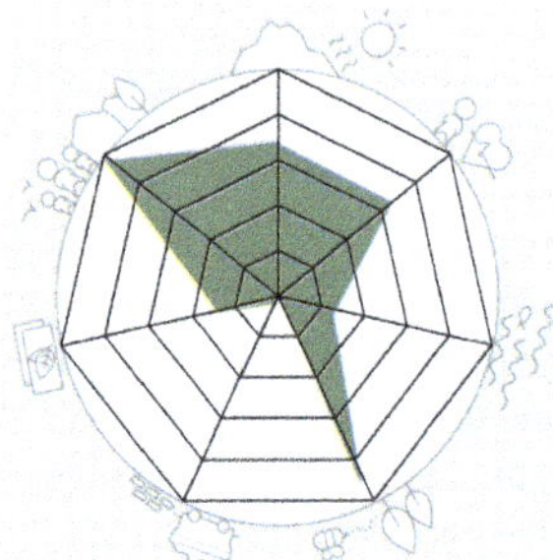

**Type**
Functional

**Building part**
Outdoor equipment

**Impact category**
Community

**Material**
Vegetation a/o water

**Intervention**
N.D.

**Scale**
Site

**Strategy**
Ecosystem protection
Community engagement

**Description**

Integrate green areas in the service of employees and the community by implementing:
i) organic water treatment and collection systems that produce both landscape and environmental effects (e.g., ponds or water channels that characterize the spaces and improve the summer micro climate);
ii) the design of greenery with a function that is not only aesthetic but also regulates the climate (e.g., trees shielding the view but also prevailing winds and direct solar radiation);
iii) the installation of quality street furniture, lighting, and fences (O.P.U.Q.P., 2015).
Beyond designing high-quality green spaces, it is also necessary to maintain them properly. So, develop a specific green area maintenance plan, and devise mixed economic management models (Busquets i Fabregas, 2007).
If space is limited, especially in regeneration projects, a "pocket park" might be created by turning an empty interstitial area into green space to benefit the local micro climate (Comune di Bomporto, 2016).

**Practices**

a) public park for the neighbourhood
A2 - Bio Pastificio IRIS (IT, 2015)
A8 - Pastificio Felicetti (IT, 2014)

# PA 4.4 VISIBILITY AND IDENTITY OF OPEN SPACE

**Type**
Formal

**Building part**
Outdoor equipment

**Impact category**
Community

**Material**
Vegetation a/o water

**Intervention**
N.D.

**Scale**
Site

**Strategy**
Well-being of workers

Brand image

**Description**

**Improve open spaces by differentiating and making them** visible from outside the industrial area, making them appealing not only to employees but also to the community and visitors.

Create cuts and gaps in the building from open spaces to and from exterior visual centres to accomplish this, and give the core public places a distinct identity.

Then, in the most visible spots, install public equipment and create appropriate pathways (including shaded ones) for the use of green spaces (O.P.U.Q.P., 2015).

**Practices**

N.D.

# PA 5.1 SERVICES FOR WORKERS

**Type**
Functional

**Building part**
N.D.

**Impact category**
Health and Well-being

**Material**
N.D.

**Intervention**
New construction

**Scale**
Site
Building

**Strategy**
Well-being of workers

**Description**

Taking care of the quality and liveability of workplaces means creating the collective spaces required by standards, with a special emphasis on the establishment of shaded areas and walks that can be utilized for recreation and relaxation (O.P.U.Q.P., 2015).
However, it also entails providing employees with services and relational spaces that go beyond standards, such as relaxation rooms, terraces, shared and creative work areas, and sports equipment (Cassatella e Gambino, 2013; Herrmann et al., 2015).
In fact, several studies show that smart space design improves employee well-being and, as a result, productivity (Macmillan, 2006).
The strategy is especially important in the offices, where it is necessary to stimulate creativity, communication, teamwork, and the removal of hierarchies, as well as to improve employee psycho-physical health. If the company is too small to finance investments and worker spaces, shared services can be formed with neighbouring businesses.

**Practices**

a) relax area, canteen, multi-purpose rooms, square as open space for employees
A1 - Melinda (IT, 2014)
A2 - Bio Pastificio IRIS (IT, 2015)

b) different spaces for diverse needs, including spaces for informal meetings
B7 - Carlsberg (FR, 2014)

c) mixed spaces for employees and visitors
B8 - San Pellegrino (IT, ongoing)

d) roof terrace
B15 - CE.VI.V. (IT, 2018)

# PA 5.2 ERGONOMIC WORKPLACE

**Type**
Functional

**Building part**
N.D.

**Impact category**
Health and Well-being

**Material**
Coating and paint

**Intervention**
N.D.

**Scale**
Building

**Strategy**
Well-being of workers

**Description**

Carefully design workspaces, especially in offices where it is vital to encourage creativity, communication, teamwork, and the removal of hierarchies. This in fact can have a significant impact on employee well-being and, as a result, productivity (Macmillan, 2006 ).

To this end, it is critical to use different colours in work spaces (both offices and production) that can induce or inhibit staff actions: for example, the artist F. von Garnier - author of some famous chromatic redevelopments on German territory - intentionally uses pink in parts that must not be touched because they are dangerous, because workers (in those cases only men) have a natural aversion to this shade (Spagnoli, 2010).

Similarly, Jiang et al. (2019) notes that relaxing colours are associated with lower levels of stress in employees: in general, these are colours drawn from the natural environment, such as green and blue. So even in the work setting it is important to create visual distraction to relax eyes and mind.

User-controlled workstations can also be important because people's needs change and even the same person may have different needs and preferences at different times. As a result, the ability to adjust the indoor atmosphere is an important factor in occupant comfort and satisfaction (U.S. GBC, 2014). This includes temperature, humidity, and air flow, which can all be controlled, for example, via operable windows. This, however, may be in conflict with the energy efficiency of the building, so it must be checked with the overall system requirement.

**Practices**

a) space is designed to give the surveyor good visibility while also making the visitors feel at ease
B6 - Dalmunach distillery (UK, 2015)

b) colours and shapes for workers' safety and well-being
B7 - Carlsberg (FR, 2014)

# PA 5.3 FLEXIBLE SPACE

**Type**
Functional

**Building part**
Indoor partition

**Impact category**
Material resources

**Material**
N.D.

**Intervention**
New construction

**Scale**
Building

**Strategy**
New material saving
Flexibility

**Description**

Creating flexible, free, and open work settings without pre-defined hierarchies can enhance worker productivity, cooperation, and efficiency, particularly in administrative contexts.

Flexibility in production spaces, on the other hand, provides for the extension of available space for production or storage based on needs, as well as the conversion of space to other activities at specified periods or seasons of the year.

When production is halted in the evenings or during particular seasons, for example, the scale of the industrial facilities provides for a variety of leisure uses (e.g., shows, cinemas, dance schools).

**Practices**

a) productive space in winter, they are adapted for community uses for the rest of the year (cinema, events)
A15 - Brown Sugar Factory (CH, 2016)

b) part of the canteen and education centre can be separated from the whole and open at night
A21 - Van Gelder (NL, 2019)

c) connection spaces among major buildings are flexible and create alternatively porch and corridors or workshops
A27 - Tangshan Organic (CH, 2016)

# PA 6.1 CONSIDERATION OF THE LANDSCAPE'S VALUE

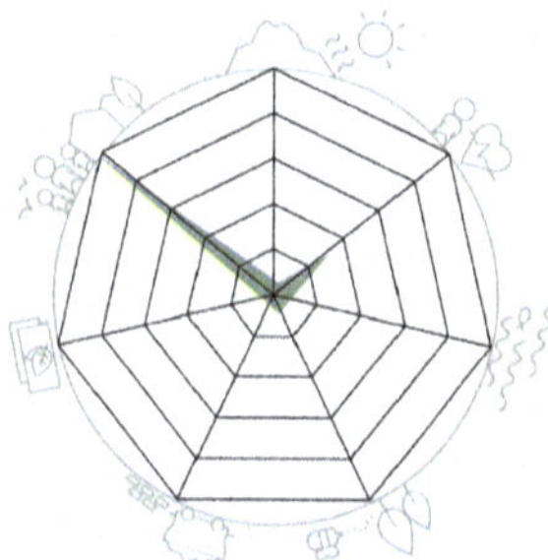

**Description**

To integrate with the local socio-cultural component, respect the preferences, associations, and memories of populations that have settled in a territory.

This involves, for example, understanding the place's archetypal forms, dominating colours, materials, and identity/iconic sense.

Active participation of stakeholders in the design process can be very beneficial in terms of establishing a solid knowledge foundation.

**Type**
Formal

**Building part**
N.D.

**Impact category**
Community

**Material**
N.D.

**Intervention**
N.D.

**Scale**
Site

Building

**Strategy**
Visual mitigation

**Practices**

a) colours and materials typical of the countryside
A3 - Almazara Olisur (CL, 2009)

b) underground building covered by fruit trees and cultivated fields typical of the context
A9 - Celle ipogee Melinda (IT, 2004)

c) new landmark that respects the natural scenery
A14 - Tofu Factory (CH, 2018)

d) forms and materials in continuity with the context
A26 - Fratelli Beretta (U.S., 2016)
B2 - Cantina Pizzolato (IT, 2016)
B3 - Cantina Antinori (IT, 2013)

e) respect of the environment and the scene as mirrored in the cooperative values
A28 - Mulino Agri '90 (IT, 2011)

f) harmonisation with shape, materials and symbols of the landscape
B5 - Macallan distillery (UK, 2018)

g) respect for the iconic value of the previous building, demolished only in part as necessary
B17 - Distilleria Zanin (IT, 2013)

# PA 6.2 PERSPECTIVE AND VISUAL CHANNELS

**Type**
Formal

**Building part**
N.D.

**Impact category**
Community

**Material**
N.A.

**Intervention**
New construction

**Scale**
Site

**Strategy**
Visual comfort
Physical link with the context

**Description**

Respect the environment by conserving the visual cones while also boosting the production site perspective towards scenic places.

The strategy is based on detailed scenic analyses and effect assessments in order to prevent the visual impact of new industrial complexes on sensitive receptors, as well as the visual obstruction of the landscape caused by seamless curtains.

This can be attained, for example, by relocating buildings in relation to the front road, favouring in-depth settlement logic, or introducing cuts and gaps to break the wall continuity (O.P.U.Q.P., 2015).

**Practices**

N.D.

# PA 7.1 CONSISTENCY WITH BRANDING

**Type**
Formal

**Building part**
Vertical envelope

**Impact category**
Green economy

**Material**
Coating and paint

**Intervention**
N.D.

**Scale**
Building

**Strategy**
Brand image

**Description**

Communicate the company's activity through the architecture of the factory. This can improve the corporate image and increase its advertising potential, compared to an impersonal and monotonous facade treatment.
This entails, for example, making a symbolic allusion to the shape of the flagship product's packaging, the packaging materials, the product itself, or even a specific point in the manufacturing process (e.g. evaporation, freezing).
Even the creative insertion of the corporate logo onto the facade, or the recall of colours and icons familiar to consumers, can make it immediately recognized and help people passing in identifying the activities taking place inside, thereby facilitating consumers in locating the organization geographically.
The site qualification, on the other hand, cannot be limited to the affixing of plaques and advertising to the outside. So, be careful not to use advertising as a means to an end.

**Practices**

a) the colour of the building is the same of the logo, and Y-shape pillars recall the tree/vineyard
A1 - Melinda (IT, 2014); B15 - CE.VI.V. (IT, 2018)

b) a gigantic dog and cat on the roof attract visitors from the highway A20 - Harim Pet Food (KR, 2016)

c) a metallic sculpture recalls the coffee scent coming up from the cup, and Corten recalls the colour of toasted coffee A22 - Caffè Milani (IT, 2017)

d) hanging roof shields the building and recalls the packaging A4 - Stabilimento Maina (IT, 2015)
A12 - Bovetti (FR, 2010)

e) the shape of greenhouses recall the steam of the process, and contain two plant families used to flavour gin
B14 - Bombay Sapphire Distillery (UK, 2013)

# PA 7.2 REFERENCE TO PRODUCT'S INGREDIENTS

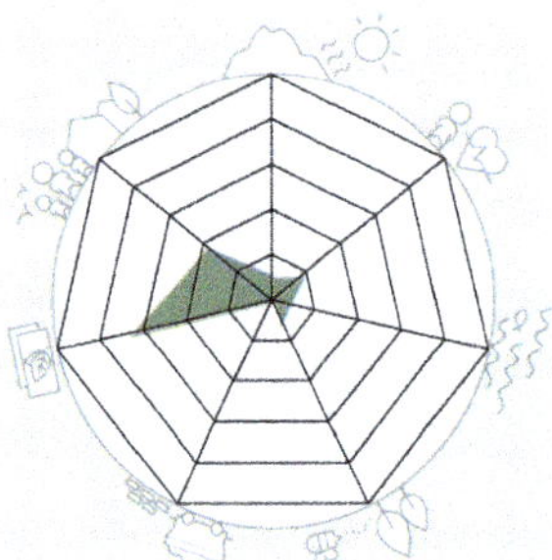

**Type**
Formal

**Building part**
Outdoor equipment

**Impact category**
Green economy

**Material**
Vegetation a/o water

**Intervention**
N.D.

**Scale**
Site

**Strategy**
Brand image

**Description**

Recall the basic ingredients of products or manufacturing processes can be especially useful in complementing and characterizing open spaces.

For example, in the food industry, this strategy enables the recall of water and vegetable crops that are at the heart of many processes, transforming them into high-quality elements (e.g., ponds, small water channels, gardens) that decorate the area surrounding the factory and serve as a pretext for educational visits to the site.

**Practices**

a) coffee plants are cultivated in a greenhouse with educational purposes
A22 - Caffè Milani (IT, 2017)

b) helical staircase as a curl of the grape and vineyard on the roofing
B3 - Cantina Antinori (IT, 2013)

c) the basic ingredients of the process feature open spaces
B5 - Macallan distillery (UK, 2018)
B6 - Dalmunach distillery (UK, 2015)
B7 - Carlsberg (FR, 2014)

# PA 7.3 SHOWCASE OF TECHNOLOGY

**Type**
Formal

**Building part**
Systems

**Impact category**
Climate change

**Material**
N.D.

**Intervention**
N.D.

**Scale**
Site

Building

**Strategy**
Energy efficiency

Brand image

**Description**

Display rather than hide the technical devices that characterize the production or that are installed to achieve technological performance above the standard to boost **their relevance and use them as an identification element** of the project.

The tactic is frequently used in the so-called "zero emission factories," which display as formal elements the technological and plant engineering devices used to achieve their exceptional environmental performances. For example, the presence of photovoltaic panels or rainwater collection tanks/silos is often highlighted.

**Such practices are especially significant because** communicating a company's environmental behaviour is **now a critical feature of a firm's competitive positioning** (Cavallo, Degli Esposti, Konstantinou, 2012).

**Practices**

a) photovoltaic panels as formal elements
A5 - Grandi Salumifici Italiani (IT, 2013)

b) showcasing of silos and water tanks as formal elements
B1 - UC Davis Wine Brewery (CA, 2013)
B4 - Bodegas Portia (ES, 2010)

# PA 7.4 BUILDING AS SHOWCASE OF THE PRODUCT

**Type**
Formal

**Building part**
Vertical envelope

**Impact category**
Green economy

**Material**
Glass

**Intervention**
N.D.

**Scale**
Building

**Strategy**
Green marketing

**Description**

Create a building, or parts of a building, that exhibit the company's products from the outside, as if it were a large shop window.
The architecture of the factory thus becomes a source of communication and promotion for the corporation, based on the principle of transparency and consumer openness.

**Practices**

a) ingredients (crops) hung from the roof to provide sun protection while drying
A18 - Iroha Village Factory (J, 2018)

b) products (hams) hung in front of the glass wall
A19 - Opificio Salpi (IT, 2010)

c) as a showcase the glass building make the process visible, and ingredients (fruit trees) are planted in the park
A21 - Van Gelder (NL, 2019)

d) a large glazing shows silos containing beer
B12 - Bavaria Brewery (CO, 2011)

# PA 7.5 CONSISTENCY WITH COMPANY VALUES

**Type**
Management

**Building part**
N.D.

**Impact category**
Green economy

**Material**
N.D.

**Intervention**
N.D.

**Scale**
Site

Building

**Strategy**
Green marketing

**Description**

A principle of coherence between the architecture of the industrial complex and corporate ideals must be pursued in order to promote and embody company values in the architecture.

The use of factory itself as a form of communication is not new. However, it is increasingly being taken to bad extremes, sometimes reducing the building to a mere media support with the risk of the workplace becoming "Disneyfied" (Qualizza, 2012).

Avoid green washing processes and instead completely integrate the company's environmental sustainability concepts at the factory level by adopting efficient and environmentally friendly construction methods and plant technology.

**Practices**

a) respect for the environment: green factory
A2 - Bio Pastificio IRIS (IT, 2015)

b) innovation: energy demand covered by photovoltaics
A5 - Grandi Salumifici Italiani (IT, 2013)

c) innovation: communication of the technological content
A10 - Pellini Caffè (IT, 2006)

d) narration: the façade illustrates the company's values A8 - Pastificio Felicetti (IT, 2014)

e) territorial promotion: food storage underground to leave the cultivated landscape intact
A9 - Celle ipogee Melinda (IT, 2004)

f) hygiene: clean and secure food is produced in spaces where materials are highly washable, clear and smooth
A20 - Harim Pet Food (KR, 2016)

g) respect for the environment: energy retrofitting and low life cycle impact materials
A25 - Felsineo Veg (IT, 2017)

# PA 7.6 GREEN BUILDING CERTIFICATION

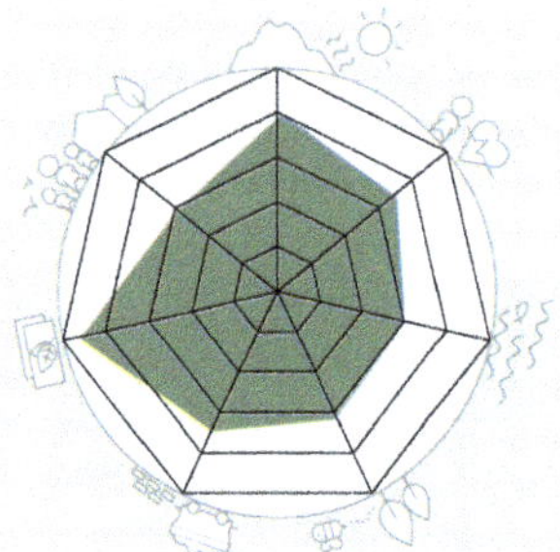

**Type**
Management

**Building part**
N.A.

**Impact category**
Green economy

**Material**
N.A.

**Intervention**
N.D.

**Scale**
Site

Building

**Strategy**
Energy efficiency

Green marketing

**Description**

Certify a production site's environmental performance in order to raise its level and communicate the degree of sustainability to the outside world, so boosting its reputation based on actual data and numbers.

In fact, according to ANIT (2012), environmental **sustainability certification is a tool that allows a building's** performance and environmental impact on the area to be properly acknowledged.

More and more businesses are embracing voluntary tools for assessing building sustainability. These have often the additional goal of guiding the design stage with an eye toward environmental, social, and economic sustainability, and thus also connecting the physical characteristics of buildings to a concept consistent with the company key values (Qualizza, 2012).

LEED, DGNB, BREEAM, CASBEE, and Protocollo ITACA are some of the most well-known Green Building Rating Systems (GBRSs) relevant to industries.

**Practices**

a) DGNB
A6 - Solare Manufaktur Peter (DE, 2012)

b) LEED
A16 - Mars Petcare (AU, 2015)
A17 - Frito-Lay (TX, 2011)
B1 - UC Davis Wine Brewery (CA, 2013)
B11 - Valprè (ZA, 2011)

c) BREEAM
A21 - Van Gelder (NL, 2019)
B14 - Bombay Sapphire Distillery (UK, 2013)

d) combination: LEED Platinum + BREEAM Outstanding
B13 - St. James Gate Brewery (UK, 2011)

# PA 7.7 EXPERIENCES FOR CONSUMERS

**Description**

Rethink the factory as a "place" rather than a "space" (Qualizza, 2012): a place where the productive process is open to the visitor in certain parts, featured by transparent and spectacular aspects.

The tactic is especially relevant in the food and beverage industry, which is increasingly subjected to food and wine tourism phenomena. Visitors want to experience the excellence of the place and process from which the wine, bread, or beer they purchase is made.

**Type**
Management

**Building part**
N.D.

**Impact category**
Green economy

**Material**
Glass

**Intervention**
N.D.

**Scale**
Site

Building

**Strategy**
Stakeholder engagement

Green marketing

**Practices**

a) visitors' pathway with glass walls and windows to see the process
A14 - Tofu Factory (CH, 2018)

b) glass wall (transparent) at ground level for visitor to see the process
A15 - Brown Sugar Factory (CH, 2016)

c) guided tours of the factory
A18 - Iroha Village Factory (J, 2018)

d) animals and their owners can see the process through an ad-hoc window "where is my food"
A20 - Harim Pet Food (KR, 2016)

e) greenhouse on the roof as visitors' attraction after completing the tour of the process
A21 - Van Gelder (NL, 2019)

f) views on the process, space and paths for food and wine tourism
B3 - Cantina Antinori (IT, 2013)
B4 - Bodegas Portia (ES, 2010)
B5 - Macallan distillery (UK, 2018)
B6 - Dalmunach distillery (UK, 2015)
B14 - Bombay Sapphire Distillery (UK, 2013)
B16 - Roe & Co Distillery (IE, 2019)

# PA 7.8 DYNAMIC IMAGE ON THE ROAD

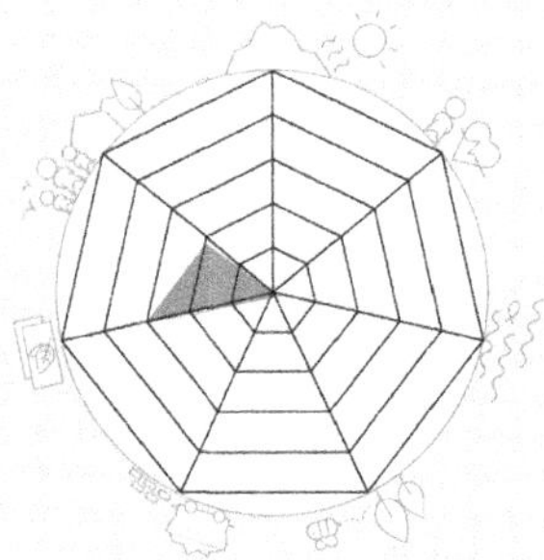

**Type**
Formal

**Building part**
Vertical envelope

**Impact category**
Green economy

**Material**
N.D.

**Intervention**
N.D.

**Scale**
Site
Building

**Strategy**
Brand image

**Description**

Consider the location on major roads as an opportunity in terms of communication.

In this case, focusing on the image of movement, on the strong and identifying sign in the motorway landscape, **where the rapid and effective perception of flashes of** colour, symbols, and recognizable - or memorable - icons play a fundamental role in communication and promotion of the corporate image through the architecture, can be **beneficial.**

Because perception is so fast, there is no time for details. Rather the tactic consists in the use of macro systems, such **as large fields of colour, or accentuated profiles that recall** the principles of the company, or even games of light that are activated in the evening.

As a result, the company's location on a high-speed route **may become a strength and a significant aspect of its** marketing strategy.

**Practices**

a) singularity: positive contrast between the building and the surrounding
A4 - Stabilimento Maina (IT, 2015)

b) façade restyling on the highway to communicate technological innovation
A5 - Grandi Salumifici Italiani (IT, 2013)

c) façade restyling on the highway resulting in a landscape landmark
A23 - Carnitalia (IT, 2016)

d) the building skyline recalls the company logo
B7 - Carlsberg (FR, 2014)

# PA 7.9 PROCESS COMMUNICATION OUTWARDS

**Type**
Formal

**Building part**
N.D.

**Impact category**
Green economy

**Material**
N.D.

**Intervention**
New construction

**Scale**
Building

**Strategy**
Green marketing

**Description**

Communicating the production process or internal operations to the outside world can help the corporation demonstrate the excellence of its processes while also informing the consumer or driver about the contents of the building.

Furthermore, if it is well-designed, it can be used with other tactics to contribute to the visual mitigation of the site.

For example, separate the site many functional parts morphologically to emphasize the various production phases; or exhibit silos and storage tanks as a didactic feature.

**Practices**

a) morphological emphasis of diverse production areas
B5 - Macallan distillery (UK, 2018)

b) silos and water tanks as didactic elements
B6 - Dalmunach distillery (UK, 2015)

c) each floor corresponds to a production phase (from the top down, the process works with gravity)
B9 - Cantina di Collemassari (IT, 2005)

## References


- ANIT. (2012). Linee guida per la progettazione con i protocolli di sostenibilità. Xella Italia.
- Aster. (2014). Le Priorità Tecnologiche Regionali. Agroalimentare.
- Ball, P. D., Despeisse, M., Evans, S., Greenough, R. M., Hope, S. B., Kerrigan, R., Levers, A., Lunt, P., Murray, V., Oates, M. R., Quincey, R., Shao, L., Waltniel, T., & Wright, A. J. (2013). Factory modelling: Combining energy modelling for buildings and production systems. IFIP Advances in Information and Communication Technology, 397(PART 1), 158–165. https://doi.org/10.1007/978-3-642-40352-1_21
- Boeri, C., Braz de Oliveira, I., & Giambruno, M. C. (2013). Colour design and industrial landscape. III CITCEM Conference.
- BRE. (2016). BREEAM International New Construction 2016. Technical Manual.
- Busquets i Fabregas, J. (Ed.). (2007). Per una corretta visione del paesaggio: linee guida (Traduzione). Generalitat de Catalunya. http://www.paysmed.net/
- Cassatella, C., & Gambino, R. (2013). Linee guida per i paesaggi industriali, un'esperienza di ricerca. XXXIV Conferenza Italiana Di Scienze Regionali, 1–14.
- Cavallo, M., Degli Esposti, P., & Konstantinou, K. (Eds.). (2012). Green Marketing per le aree industriali. Metodologie, strumenti e pratiche. Franco Angeli.
- Chowdhury, S., Hamada, Y., & Shabbir Ahmed, K. (2017). Indoor heat stress and cooling energy comparison between green roof (GR) and non-green roof (n-GR) by simulations for labor intensive factories in the tropics. International Journal of Sustainable Built Environment, 6, 449–462. https://doi.org/10.1016/j.ijsbe.2017.09.001
- Clark, J. P. (2005). Food plant design. Food engineering. In G. Barbosa-Canovas (Ed.), Enciclopedia of Life Support Systems (pp. 683–696). EOLSS Publisher/UNESCO.
- Comune di Bomporto & Consorzio Attività Produttive Aree e Servizi di Modena. (2016). Bomporto (MO) Piano di adattamento. www.lifeiris.eu
- DEFRA. (2006). Food Industry Sustainability Strategy (FISS). In Department for Environment, Food and Rural Affairs. https://www.gov.uk/government/uploads/system/uploads/attachment_data/file/69283/pb11649-fiss2006-060411.pdf
- Delpiano, A., & Trombetti, G. (Ed. . (2011). EnergiApea. Vincere la sfida energetica (RicercaeProgetto - Galassi Mingozzi e associati). Tipografia metropolitana Bologna.
- Despeisse, M. (2013). Sustainable Manufacturing Tactics and Improvement Methodology. A Structured and Systematic Approach to Identify Improvement Opportunities [School of Applied Science, Cranfield University]. https://doi.org/10.1017/CBO9781107415324.004
- Despeisse, M., Davé, A., Litos, L., Roberts, S., Ball, P., & Evans, S. (2016). A Collection of Tools for Factory Eco-efficiency. Procedia CIRP, 40, 542–546. https://doi.org/10.1016/j.procir.2016.01.130
- Dessì, V., Farnè, E., Ravanello, L., Salomoni, M. T., Dessì, V., Farnè, E., Ravanello, L., & Salomoni, M. T. (2018). Rigenerare la città con la natura (2nd ed.). Maggioli Editore.
- Duflou, J. R., Sutherland, J. W., Dornfeld, D., Herrmann, C., Jeswiet, J., Kara, S., Hauschild, M., & Kellens, K. (2012). Towards energy and resource efficient manufacturing: A processes and systems approach. CIRP Annals - Manufacturing Technology, 61, 587–609. https://doi.org/10.1016/j.cirp.2012.05.002

- García, L., Hernández, J., & Ayuga, F. (2003). Analysis of the exterior colour of agro-industrial buildings: A computer aided approach to landscape integration. Journal of Environmental Management, 69(1), 93–104. https://doi.org/10.1016/S0301-4797(03)00121-X
- Ginelli, E., & Daglio, L. (2016). Le infrastrutture per le energie rinnovabili nel paesaggio. Strumenti di progetto e traiettorie dell'innovazione. TECHNE, 11, 119–126. https://doi.org/10.13128/Techne-18410
- Grant, J. (1997). Planning and designing for eco-efficiency industrial landscapes. Journal of Cleaner Production, 5(1–2), 75–78.
- Herrmann, C., Blume, S., Kurle, D., Schmidt, C., & Thiede, S. (2015). The positive impact factory - Transition from eco-efficiency to eco-effectiveness strategies in manufacturing. Procedia CIRP, 29, 19–27. https://doi.org/10.1016/j.procir.2015.02.066
- Jiang, B., Wang, H., Larsen, L., Bao, F., Li, Z., & Pryor, M. (2019). Quality of sweatshop factory outdoor environments matters for workers' stress and anxiety: A participatory smartphone-photography survey. Journal of Environmental Psychology, 65(January), 101336. https://doi.org/10.1016/j.jenvp.2019.101336
- Kaur, G. P., Gupta, P., & Syal, M. (2016). Adoption of green practices in industrial buildings: an action research on capacity building of stakeholders towards green factories. International Journal of Sustainable Land use and Urban Planning, 3(2), 1–12.
- Koçak, B., Fernandez, A. I., & Paksoy, H. (2020). Review on sensible thermal energy storage for industrial solar applications and sustainability aspects. Solar Energy, 209(March), 135–169. https://doi.org/10.1016/j.solener.2020.08.081
- Macmillan, S. (2006). Added value of good design. Building Research & Information, 34(3), 257–271. https://doi.org/10.1080/09613210600590074
- May, G., Stahl, B., & Taisch, M. (2016). Energy management in manufacturing: Toward eco-factories of the future - A focus group study. Applied Energy, 164, 628–638. https://doi.org/10.1016/j.apenergy.2015.11.044
- Mazarrón, F. R., Cid-Falceto, J., & Cañas, I. (2012). Ground thermal inertia for energy efficient building design: A case study on food industry. Energies, 5(2), 227–242. https://doi.org/10.3390/en5020227
- Miah, J. H., Griffiths, A., McNeill, R., Poonaji, I., Martin, R., Morse, S., Yang, A., & Sadhukhan, J. (2015). Creating an environmentally sustainable food factory: a case study of the Lighthouse project at Nestlé. Procedia CIRP, 26, 229–234. https://doi.org/10.1016/j.procir.2014.07.030
- Mueller, F., Cannata, A., & Herrmann, C. (2014). Procedure of modular green factory planning to enhance collaboration and decision making. Procedia CIRP, 15, 331–336. https://doi.org/10.1016/j.procir.2014.06.071
- O.P.U.Q.P. (2015). Linee guida per i paesaggi industriali in Sardegna.
- Osservatorio GreenER. (2014). Osservatorio green economny regionale. Analisi & tendenze. Rapporto 2014.
- Philipp, M., Schumm, G., Peesel, R. H., Walmsley, T. G., Atkins, M. J., Schlosser, F., & Hesselbach, J. (2017). Optimal energy supply structures for industrial food processing sites in different countries considering energy transitions. Energy, 1–12. https://doi.org/10.1016/j.energy.2017.05.062
- Provincia di Bologna. (2008). APEA - Aree Produttive Ecologicamente Attrezzate. Linee

guida (M. Cavallo (Ed.)).

- Provincia di Bologna - Servizio Artigianato Commercio e Industria. (2009). Risparmio energetico negli edifici ad uso produttivo (pp. 1–122).
- Provincia di Milano. Adeguamento del PTCP alla LR 12/2005. Repertorio delle misure di mitigazione e compensazione paesistico-ambientali, 1 (2012).
- Qualizza, G. (2012). Inventare lo spazio di una green factory. In M. Cavallo, P. Degli Esposti, & K. Konstantinou (Eds.), Green marketing per le aree industriali. Metodologie, strumenti e pratiche (pp. 220–238). Franco Angeli.
- Reichardt, J. (2015a). Master Building Plan. In H.-P. Wiendahl, J. Reichardt, & P. Nyhuis (Eds.), Handbook Factory Planning and Design (pp. 311–330). Springer.
- Reichardt, J. (2015b). Site Planning from Space View. In H.-P. Wiendahl, J. Reichardt, & P. Nyhuis (Eds.), Handbook Factory Planning and Design (pp. 331–339). Springer.
- Simeoni, P., Ciotti, G., Cottes, M., & Meneghetti, A. (2019). Integrating industrial waste heat recovery into sustainable smart energy systems. Energy, 175, 941–951. https://doi.org/10.1016/j.energy.2019.03.104
- Spagnoli, M. (2010). Friedrich Ernst von Garnier: un pioniere del color design Friedrich / Ernst v. Garnier: a true color design pioneer. In IDC Colour Centre, C. Boeri (Ed.), Colore. Quaderni di cultura e progetto del colore.
- Tandy, C. (1975). Landscape of industry. Leonard Hill Books.
- U.S. GBC. (2014). LEED Core Concepts Guide. U.S. Green Building Council.